The Konstantinos Karamanlis Institute for Democracy Yearbook Series

Series Editor:
Konstantina E. Botsiou
Associate Professor of Modern History and International Politics at the University of Peloponnese, Corinth, Greece
Director General, Konstantinos Karamanlis Institute for Democracy

The Konstantinos Karamanlis Institute for Democracy
10, Vas. Sofias Ave.
10674 Athens
Greece
www.idkaramanlis.gr, info@idkaramanlis.gr

The Konstantinos Karamanlis Institute for Democracy Yearbook
series features collections of essays on international politics, written from a European point of view. Each volume reflects on events that marked the previous year and addresses the challenges ahead. Eminent political figures, academics, diplomats, journalists, and professionals offer their views on diverse political, economic, social, and ideological issues that have shaped and continue to affect contemporary political and social dynamics within and beyond state borders.

Konstantina E. Botsiou • Antonis Klapsis
Editors

The Konstantinos Karamanlis Institute for Democracy Yearbook 2011

The Global Economic Crisis and the Case of Greece

Editors
Dr Konstantina E. Botsiou
Associate Professor, University of Peloponnese
Director General
Konstantinos Karamanlis Institute
for Democracy
10, Vas. Sofias Ave.
10674 Athens
Greece
kbotsiou@idkaramanlis.gr

Dr Antonis Klapsis
Head of Publications and Research
Programmes
Konstantinos Karamanlis Institute
for Democracy
10, Vas. Sofias Ave.
10674 Athens
Greece
klapsis@idkaramanlis.gr

This is a joint publication of the Centre for European Studies and the Konstantinos Karamanlis Institute for Democracy. This publication receives funding from the European Parliament.

ISSN 1868-2111 e-ISSN 1868-212X
ISBN 978-3-642-18414-7 e-ISBN 978-3-642-18415-4
DOI 10.1007/ 978-3-642-18415-4
Springer Heidelberg Dordrecht London New York

The Konstantinos Karamanlis Institute for Democracy Yearbook Series

The processing of the manuscript was concluded in 2010

Cover design: eStudio Calamar S.L.

Printed on acid-free paper

Springer is part of Springer Science+Business Media (www.springer.com)

Acknowledgements

Many people have contributed great effort and commitment to bringing this volume to publication. First of all, we would like to express our gratitude to the Centre for European Studies in Brussels for its valuable collaboration in completing this joint venture. We are indebted to Katharina Wetzel-Vandai and Irene Barrios-Kežić at Springer for the support and encouragement they offered to this project. We also thank the Communicative English editing team for their meticulous copyediting of the draft manuscript. We are grateful to Antigoni Kouvidi at the Department of Publications of the Konstantinos Karamanlis Institute for Democracy for elaborating and editing the draft texts. Finally, we are deeply indebted to the scholars and policymakers whose work is included in this volume for their participation and contributions.

Foreword

The year 2011 will likely be a critical turning point for Europe. Many European countries, including Greece, are experiencing the negative effects of an unprecedented economic crisis. In fact, for the first time in decades a substantial number of European Union members are facing serious financial distress, which in some cases might even result in national insolvency. Above all, the crisis threatens to undermine the foundations of social cohesion, which in turn could trigger political instability. Thus, the crucial question to be answered is this: Will the EU and its member states manage to overcome the great obstacles in their way, or will the difficulties prove fatal for European integration?

A proper answer cannot be given without first taking into consideration the very essence of politics, which in democracies is closely associated with collective decision-making. It follows that it is the obligation of political leaders, at both the national and the supranational level, to present to their fellow citizens suggestions that can provide realistic and viable solutions to the great economic problems. In this context, radically reducing public deficits at any cost is not the best approach to take, as it will probably lead to the creation of a vicious circle of economic depression with unpredictable social consequences. In other words, contractionary fiscal measures should be coupled with initiatives that boost economic development and productivity and that can heal the real source of the problem.

At the same time, a new form of economic governance at the European level is absolutely necessary so that the EU will be in a position to deal with even more complex problems in the future. Solidarity should be the cornerstone of such governance, while the concept of the welfare state should remain the core of any future policy. It is, after all, obvious that the quest for prosperity is better served through collective action and mutual assistance – this was the opinion of the founders of European integration. As Jean Monnet put it, "There is no future for the people of Europe other than in union."

These and other challenges are reflected in the pages of the *Konstantinos Karamanlis Institute for Democracy Yearbook 2011*. The analyses by prominent scholars and policymakers aim to shed light on different aspects of the complex Greek and international agenda as well as contribute to public debate. It is hoped that the variety of national and academic backgrounds represented by the authors will stimulate further discussion on democratic governance and European integration, thus fulfilling one of the Institute's main goals.

Evangelos Meimarakis
President of the Konstantinos Karamanlis Institute for Democracy
MP, former Minister of Defence of Greece

Contents

Contributors

Dr Robert Z. Aliber
Emeritus Professor of International Economics and Finance, Booth Graduate School of Business at the University of Chicago

Dr Leszek Balcerowicz
Professor at the Warsaw School of Economics

Dr Christos Dimas
PhD in European Politics and Policy, the London School of Economics and Political Science

Dr Ross Fakiolas
Emeritus Professor of Economics at the National Technical University of Athens

Dr Christos Gortsos
Associate Professor of International Economic Law in the Department of International and European Studies at the Panteion University of Athens, Visiting Professor at the Law School of the University of Athens and at the Europa Institute, University of Saarland

Dr Anthony Ioannidis
Assistant Professor of Management in the Department of Business Administration at the Athens University of Economics and Business

Dr Emmanuel Kakaras
Professor of Thermal Power Generation in the School of Mechanical Engineering at the National Technical University of Athens

Sir Michael Llewellyn Smith
British Ambassador to Greece, 1996–9, and historian

Dr Helene Mandalenakis
PhD in Political Science and International Relations, McGill University, Canada, and Adjunct Lecturer in the Department of Political Science and International Relations at the University of Peloponnese

Dr Ioannis A. Mourmouras
Professor of Macroeconomics in the Department of Economics at the University of Macedonia, Thessaloniki, and Advisor to the Leader of the Opposition and President of the Nea Demokratia Party, Mr Antonis Samaras

Dr Harris Mylonas
Assistant Professor of Political Science and International Affairs, The Elliot School of International Affairs, George Washington University, and Academy Scholar, Harvard Academy for International and Area Studies, Harvard University

Dr Pyrros Papadimitriou
Lawyer and Economist, Assistant Professor of International Economic Relations in the Department of Political Science and International Relations at the University of Peloponnese

Dr E. S. Savas
Presidential Professor of Public Affairs at the School of Public Affairs, Baruch College, City University of New York

Dr Pantelis Sklias
Associate Professor of International Political Economy in the Department of Political Science and International Relations at the University of Peloponnese

Dr Christos Zerefos
Professor of Atmospheric Physics in the Department of Geology and Geoenvironment at the National and Kapodistrian University of Athens, and Member of the Academy of Athens

Dr Antonia Zervaki
Adjunct Lecturer in the Department of Political Science and International Relations at the University of Peloponnese, former Special Advisor on EU Integrated Maritime Policy at the Academic Centre of Analysis and Planning, Hellenic Ministry of Foreign Affairs

Abbreviations

AFS	Automatic Fiscal Stabilisers
BBC	British Broadcasting Corporation
CCS	Carbon Capture and Storage
CDE	Carbon Dioxide Equivalent
CDS	Credit Default Swap
CEPR	Center for Economic Policy Research
CFP	Common Fisheries Policy
CNN	Cable News Network
CPI	Consumer Price Index
EC	European Commission
ECB	European Central Bank
EDF	Électricité de France
EEA	European Environment Agency
EEC	European Economic Community
EERP	European Economic Recovery Plan
EIP	Excessive Imbalance Procedure
EMU	Economic and Monetary Union
ESM	European Stabilisation Mechanism
EU	European Union
EU ETS	European Union Emissions Trading Scheme
GDP	Gross Domestic Product
GHG	Greenhouse Gas
GMES	Global Monitoring of the Environment and Security
ICZM	Integrated Coastal Zone Management
ILO	International Labour Organization
IMF	International Monetary Fund
IMO	International Maritime Organization
IOM	International Organization for Migration
IPCC	Intergovernmental Panel on Climate Change

IPPC	Integrated Pollution Prevention and Control
IUU	Illegal, Unreported and Unregulated
KEPE	Centre of Planning and Economic Research
NAP	National Allocation Plan
NATO	North Atlantic Treaty Organization
NBER	National Bureau of Economic Research
ND	Nea Demokratia
NER	New Entrant Reserve
NGO	Non-governmental Organisation
OCA	Optimal Currency Areas
OECD	Organisation for Economic Co-operation and Development
OTE	Hellenic Telecommunications Organization
PASOK	Panhellenic Socialist Movement
PIB	Public Investments Budget
PPP	Public–Private Partnership
R&D	Research and Development
RES	Renewable Energy Sources
SDRM	Sovereign Debt Restructuring Mechanism
SET Plan	European Strategic Energy Technology Plan
SGP	Stability and Growth Pact
SPV	Special Purpose Vehicle
TFEU	Treaty on the Functioning of the European Union
UK	United Kingdom
UNCLOS	United Nations Convention on the Law of the Sea
UNDP	United Nations Development Programme
UNFCCC	United Nations Framework Convention on Climate Change
US	United States
VAT	Value Added Tax
WHO	World Health Organization

Introduction

Konstantina E. Botsiou and Antonis Klapsis

In times of crisis the classical linkage between leadership and reform emerges with a renewed relevance. This issue has gained particular importance in the European Union due to its peculiar supranational consensus culture. From the inception of the Communities until the current global economic crisis, the history of integration offers abundant examples of national and EU-wide structural readjustments that have brought to the fore exceptional leadership.

Apart from the very essence of integration, which has dramatically changed the conduct of national politics, the political-economic mix of the European welfare state was meant to close the door not only to communism but also to laissez-faire capitalism, which was held largely responsible for the Crash of 1929. Evidently, the profound crisis caused by war and devastation enabled far-reaching reform on the national level, too, by European statesmen like Konrad Adenauer, Robert Schuman, Alcide de Gasperi and others. A deep recession generated Eurosclerosis in the 1970s, but at the same time prepared the ground for a new social contract in the European Communities, hence triggering their transformation into the European Union, with a monetary union at its core.

The revision of the European Treaties provided a long-lasting agenda-setting reform aimed primarily at promoting the EU's international economic competitiveness. This reform collided with the expected predominance of the US for quite a few years. In the past two years, however, the US has been in the grip of a bitter economic decline, having to deal with fast-growing competitors from Asia such as China and India. The recurring question regarding the economic sustainability of the euro now involves a further, crucial issue of whether the EU is sustainable in the absence of deeper political integration. The avalanche of national debt crises among member states in the past year has provoked numerous and often mutually contradicting interpretations of the defects that a common currency suffers when it includes heterogeneous national economies.

The need for fair, strict and strong economic governance is broadly recognised. What is still missing is the balance between supranational and intergovernmental authority in what will ultimately be a wiser economic architecture. Traditionally, intergovernmental cooperation is more effective and convincing when a crisis dictates commitment and fast decision-making. Even the Delors Commission, a symbol of supranational political achievement in EU history, was stimulated by strategic choices made by powerful heads of states and governments, who agreed on the "leap forward". From this perspective, the EU summits of 2011 could have a lasting effect on the economic and political organisation of both the European Union as well as many of its member states.

Greece, Ireland, Portugal and Spain will certainly be faced with more serious social difficulties in their efforts to link economic stabilisation with drastic domestic political reform. However, the general situation is anything but easy for the stronger members, as they cannot solve single-handedly the problems generated by the common project of the euro. To be sure, national debt crises will provide fuel for heated ideological and political debates as to the overarching cause and identity of the European Union.

With this context in mind, we have dedicated a special section in the *Yearbook 2011* to addressing the global economic crisis and the case of Greece. In other essays, issues of leadership and governance point to the synergies between economic and political transformation. Finally, a few selected contributions offer updates on the major global issue of climate change, an issue closely intertwined with growth strategies even though its significance is often overtaken by more narrowly defined economic priorities.

To cope with the economic crisis, Greece has been tied to a long-term programme of fiscal austerity that is meant to undo protectionist and consumerist practices established in the course of the last 30 years. In a full reversal of its pro-spending electoral campaign, the Socialist government that was elected in 2009 resorted to a long-term programme of fiscal austerity, supervised through a novel mechanism set up by the International Monetary Fund, the European Commission and the European Central Bank in the spring of 2010. Overwhelming waves of long-range, painful reforms are being introduced on the basis of technocratic IMF-EU advice rather than arising from home-grown political planning. As radical cuts in salaries and pensions are implemented horizontally while the tax system remains largely unreformed, broad segments of Greek society are experiencing unfair allocations of the financial burden and are joining the already widespread social unrest. The profound deterioration in the standard of living is coupled with a lack of perspective as to the new social contract that will be born when the Greek economy unfreezes again. This widespread uncertainty is vividly reflected in the growing tendency of the younger generation to pursue educational and employment opportunities in other European countries.

Greece will not be the only EU country to seek a future through IMF-EU support and control. Ireland has already negotiated a parallel solution, while Portugal and Spain have proven vulnerable to the effects of the economic crisis. Rebalancing the impact of the euro on the various Eurozone economies through a fairer structure of rights and obligations that will be carefully monitored will be a great challenge for the EU in 2011.

This volume features analyses of important aspects of the above-mentioned themes. Ioannis A. Mourmouras explains how Greece could exit the economic deadlock with a domestically driven reform that avoids the predictable vicious cycle of deficits and recession. Leszek Balcerowicz examines the crisis in the Eurozone and the instruments developed by the EU to remedy precarious economies. Pyrros Papadimitriou explores fiscal rules and procedures that can stabilise the fiscal situation in the Eurozone. Helene Mandalenakis analyses the effects of the economic crisis on international migration, which can cause explosive conditions in Western societies if left to grow uncontrollably. Ross Fakiolas highlights the structural defects of the Greek economy that have surfaced under the pressure of its debt and fiscal problems. Pantelis Sklias underlines the contribution of the European and global political-economic context to national economic failures, especially in regard to the complexity and political fragility of EU institutions. Harris Mylonas investigates "the deep and proximate causes" as well as the major socio-economic consequences of the economic crisis in Greece, stressing the need to raise competitiveness by fighting deep-rooted disincentives such as nepotism, clientelism and corruption. Robert Z. Aliber sheds light on the key common features of the four waves of financial crisis that have occurred in the last 35 years, pointing out that unsustainable patterns of cash flow were largely responsible for each of these crises. Anthony Ioannidis discusses principles and versions of the Social Market Economy, the backbone of post-war European growth. Christos Gortsos examines how financial and political experience arising from the current economic crisis can be incorporated into EU economic governance so as to handle future crises more quickly and effectively than has been the case in past two years. E. S. Savas studies the role of government in generating growth and prosperity, providing several examples of minimalistic economic governance.

Crises often prompt scholars to seek analogies in the past. Revisiting history becomes popular in times of transition to new political and economic perceptions of world affairs. The issue of leadership seems central in such ventures. Sir Michael Llewellyn Smith compares the leadership qualities of three Greek statesmen from the nineteenth and twentieth centuries (Harilaos Trikoupis, Eleftherios Venizelos and Konstantinos Karamanlis) while also exploring both unique and general features of the art of leadership. Christos Dimas outlines the external reformist role of the EU in Greek politics since the 1970s.

In recent years, growth policies have increasingly included considerations of climate change. The fierce competition generated by developing economies that pay minimal attention to this issue has caused scepticism in many Western countries as to the economic viability of green and sustainable economic practices. Rising awareness of the threat of rapid climate change has not yet been fully integrated into political action.

Christos Zerefos highlights some of the most significant results of climate change, focusing on the Mediterranean area, where extreme weather phenomena reveal a rapid destabilisation of the local climate. Emmanuel Kakaras describes major EU policies for expanding renewable sources of energy over traditional ones such coal and oil. Antonia Zervaki closes the volume with an analysis of major EU policies that aim to address the thorny issue of climate change, particularly in vulnerable maritime regions, which serve as key locations for trade, communication and security.

This volume has gathered notable contributions from distinguished scholars and policymakers. We are greatly indebted to all for the excellent cooperation and valuable insights they brought into this joint project.

The Global Economic Crisis and the Case of Greece

Alternative Strategies for Greece's Exit from the Economic Crisis

Ioannis A. Mourmouras

Introduction

In early May 2010, Greece found itself on the verge of declaring a moratorium on payments as a result of inappropriate choices and omissions by consecutive governments for the past three decades, as well as the directionless governance of the country since October 2009. Greece is currently under international economic supervision and is going through the most severe economic and fiscal crisis in the post-war era. On 6 May 2010, the Socialist government signed a controversial loan agreement, the (infamous) Memorandum of Understanding, which provides for loans[1] of up to €110 billion to be lent to the country on one hand, and for the implementation of an Economic Policy Programme, entailing severe fiscal and structural measures and strict time schedules on the other. My objective here is to focus on the fiscal consolidation part of the Memorandum, going one step further. Specifically, in strictly economic terms, I compare and contrast the two alternatives presented in the public domain: on one hand, there is the Memorandum between the government and the so-called Troika – International Monetary Fund (IMF)/European Commission (EC)/European Central Bank (ECB) – and on the other hand there is the Economic Proposal of the opposition party. The stakes for our country are enormous: on one side there is the risk of a prolonged vicious cycle of deficits and recession, including an explosion of unemployment, a remarkable decline in the living standard of the Greek people, restructuring of debt or even a new

[1] These loans are characterised by strong conditionality and are granted on a quarterly basis at a 5.25% rate from Eurozone countries (2/3 of the €110 billion) and at 3.25% from the IMF.

Memorandum to be in force until 2020(!); on the other side there is the prospect of a quick economic recovery, disengagement from the Memorandum and a smooth return to international markets.

The IMF/EC/ECB–Government Memorandum: A Technical Mistake

First of all, we should consider the principal mistake in the Memorandum, which concerns the attempted fiscal consolidation, "the cornerstone of the programme" as it is distinctly referred to in the Memorandum itself: it erroneously targets the fiscal deficit instead of the structural deficit, which would have been the right approach. In other words, the cyclical deficit should *not* be a target variable for fiscal consolidation. This leads to an unprecedented launch of fiscal measures with grave consequences for the effectiveness of the overall venture. The Memorandum's reasoning is that this is indeed a vast consolidation "on paper", because many of the measures will be "absorbed" by the aggravation of the recession caused by the measures themselves. As will be shown below, both the logic of this reasoning and the effectiveness of the attempted disproportional fiscal consolidation are in question.

Every attempt at fiscal consolidation under recessionary conditions ought to take into consideration the important distinction between the cyclical deficit (caused by the economic cycle and therefore of a transitory nature) and the structural deficit (attributed to structural deficiencies and permanently affecting the overall deficit[2]). The Memorandum disregards this distinction, resulting in two major mistakes. First, when taking permanent contractionary measures to reduce the cyclical deficit, namely the deficit induced by the recession, the consequence is an even greater recession, undermining the very goal of reducing the deficit, leading to more measures – a vicious circle of deficits–austerity measures–recession. Second, much as it appears that the Memorandum takes for granted that

[2] It should be noted here that the distinction between cyclical and structural deficits is widely used in applied economic policy. Thus, apart from the fiscal consolidation programmes, this distinction also appears in the Stability and Growth Pact (as well as in the Greek SGP), in the new Economic Governance of the European Economic and Monetary Union, in the "golden rule of public finances" applied in Great Britain for many years etc.

permanent deficit reduction measures will not fully perform under recessionary conditions (for instance, a permanent increase in the VAT rate will not yield the expected revenue increase), it reacts to that reality in a completely erroneous manner.

Thus, the total number of measures provided for under the Stability and Growth Pact (SGP) in January 2010, and in the supplementary budget of March (for a measures package totalling €15 billion, or 6.8% of GDP), were increased by €5.8 billion last May, or 2.5% of GDP for the current year, with a view to reducing the deficit by 5.5% of GDP (from 13.6% in 2009 to 8.1% in 2010). For 2011, the Memorandum provides for consolidation measures of €9.1 billion (4.1% of GDP), aimed at curbing the deficit by a mere 0.5% of GDP. With respect to the period 2010–14, a consolidation of 19.7% of GDP has been planned in order to decrease the fiscal deficit by 11% of GDP! In light of the above, it is obvious to the naked eye how severe and front-loaded this fiscal consolidation in our country is. On top of that, the failure to distinguish between cyclical and structural deficits clearly demonstrates not only the disproportionate magnitude of the attempted fiscal consolidation but also its erroneous direction.

Two Impulses

Two major impulses will be created by the Memorandum. First, it will deepen and prolong the recession, thus creating a self-perpetuating vicious circle of recession and deficits; and, second, it will not be able to ensure the sustainability of our public finances – a *conditio sine qua non* for borrowing from international markets at reasonable rates – after five years (2010–14) of strict consolidation.

With respect to the first impulse, the excessive fiscal consolidation of the Memorandum is self-defeating because it will aggravate the recession (a decline of approximately 8% in the two-year period 2010–11 alone) via three distinct channels: (a) drastically reducing effective demand; (b) neutralising the effects of automatic fiscal stabilisers; and (c) affecting the expectations of the private sector. In particular, large cuts to salaries and pensions, the slashing of public investments through the Public Investments Budget (PIB) and the suspension of payments to businesses by the state on the one hand, and the "extraordinary contribution" from profitable businesses and the increase of all indirect taxes on the other hand will result in a drop in domestic demand. This drastic reduction of effective demand will lead to a self-undermining situation as regards tax collection measures: it provides for an increase of income by 13.7% for 2010, whereas this will in fact not be greater than 5%. (The list of failed predictions in the Memorandum is lengthy. I note another obvious miss: the Memorandum provides for inflation of 1.9%, whereas in 2010 it was approximately three times that figure).

The second consequence has to do with the horizontal reductions in social expenditure required by the Memorandum, which concern those who earn low wages, receive low pensions or are unemployed. Instead of attempting a rationalisation of social expenditure, the Memorandum proceeds to a reduction of pensions, allowances and other social provisions with a high multiplier, directly resulting in a deepening of the recession. Notably, the Memorandum actually neutralises the positive results of so-called automatic fiscal stabilisers.[3]

Finally, considering that the expectations of households and businesses are formed on the basis of their future income and/or wealth, as well as on general economic conditions, an excessive fiscal consolidation aggravates market expectations and psychology, which in turn – through the confidence multiplier – brings about second-round negative changes in consumption and investment expenditure and so on. Negative expectations and bad market psychology are reflected in the Economic Climate Index, which has remained at very low levels for many months now. In this light, it is no wonder that we are currently going through the greatest recession – in intensity and duration – of the last decades, one even greater than that of 1974 (-3.6%), the year of the invasion of Cyprus. Based on the Troika's own figures, in 2015 we will have reached the (real) GDP of 2008, which means that we are speaking of seven lost years in terms of national income.

[3] Automatic fiscal stabilisers (AFSs) are defined as the change in the cyclical primary fiscal balance. They are named as such due to the fact that, on one hand, they stabilise the economic cycle and, on the other hand, they are automatically generated by taxation and public expenditure (e.g., income taxes, unemployment allowances etc.). AFSs mitigate the reduction of available income and consumption expenditure of households caused by the recession (an economy where AFSs function sufficiently will have a somewhat steeper curve of aggregate product demand). Perhaps this explanation makes it easier to understand the relation between AFSs and the technical mistake of the Memorandum.

A digression may be in order here: authoritative colleagues and financial analysts alike, both in Greece and abroad, have supported the idea that the fiscal contraction alone imposed by the Memorandum involves a distinct growth aspect through so-called crowding in: the decline in interest rates and the formation of positive expectations for the private sector from future tax reductions. In fact, recent fiscal consolidations in Europe are presented in support of this assertion (e.g., Sweden, Belgium and other countries). The argument that there is a growth aspect to the Memorandum's fiscal consolidation does not apply in the case of Greece's fiscal crisis for the following reasons: (a) the European experience concerned milder fiscal consolidations and focused on reducing public expenditure, whereas a severe consolidation is currently being attempted in Greece, equally allocated between increasing taxes and reducing expenditure; (b) the international and European economic environment is currently unfavourable or at best, anemic; (c) unlike the aforementioned European countries which are highly competitive, our country has a remarkable lack of competitiveness.

The second major impasse arising from the Memorandum has to do with the agonising question of how the country will be able to borrow again from the international markets when the public debt–GDP ratio will have "climbed" from 120% in 2009 to 150% in 2014. This, in fact, is the Achilles heel of the Memorandum: after five years (2010–14) of strict fiscal consolidation, the Memorandum will still not ensure sustainability of our public finances, a *conditio sine qua non* for borrowing from international markets at reasonable rates.

There are two conditions established by the arithmetic of the dynamic government budget constraint for achieving fiscal sustainability: the creation of sufficient primary surpluses for repaying interest on debt servicing; and the achievement of high rates of growth for the de-escalation of the debt-to-GDP ratio. None of the above conditions is met, neither during the Memorandum's consolidation nor after its completion, meaning non-sustainable public finances. According to the Troika's own figures, at the end of 2014 the primary surplus (5.9% of GDP) will not suffice for interest payment (8.4% of GDP), the difference between the weighted average interest rate minus the (nominal) growth rate will be positive (+2.7), while their own projections for 10 years later (in 2020) involve a public debt in excess of 120% of GDP. According to the final draft of the 2011 budget and the Memorandum itself, namely with the figures before or after the

2009 public finances revision, debt will peak in 2013 and will decrease afterwards. However, the arithmetic of the dynamic government budget constraint is non-credible. The whole path of debt dynamics (time-profile) relies on a single number that nobody believes in, namely, that the primary surplus will rise from 3% of GDP in 2013 to 6% in 2014 and stay at that high level thereafter. The obvious question that arises is, how will this doubling of the primary surplus be achieved? Another, more pressing question is, how will the country in 2013 go to the markets and borrow (a) with a debt-to-GDP ratio over 150%; (b) with non-sustainable public debt dynamics; and (c) a heavy borrowing profile and an amortisation hump in that crucial period?

An Alternative Credible Exit Strategy

In July 2010, Greece's main opposition party, Nea Demokratia (ND), presented a thorough and complete 35-page "Proposal for Exiting the Economic Crisis", which would enable a smooth return to the markets in two years' time. The Proposal is organised in terms of targets and means, identifying as a threefold target the simultaneous tackling of recession, deficit and debt. The means for achieving this are the implementation of compensating measures of low or nil budgetary cost and the enhancement of liquidity to achieve quick economic recovery. As regards fiscal consolidation, which is undoubtedly required to achieve fiscal sustainability and be in a position to participate in international markets for borrowing at reasonable rates: with approximately half of the fiscal measures projected, the structural deficit will gradually become nil (the cyclical deficit will self-correct as the economy recovers). Through direct development of the country's "national assets", notably the immediate development of public property (chattels, real estate and intangible assets), sufficient revenue will be raised which can then be allocated for a one-off reduction in public debt.

A New Policy Mix to Tackle the Recession and the Cyclical Deficit

As stated above in relation to the problem of the measures' deficiency due to recession, the Memorandum – which is in fact three budgets, for 2010, 2011 and 2012 – provides for an excessive dosage of contractionary measures, which ultimately undermines the effectiveness of the austerity measures, aggravating the recession at the same time. Here lies the *differentia specifica* of ND's alternative Proposal: it suggests a different economic policy mix to tackle the recession caused by fiscal adjustment. What is worth noting is that this new mix tackles at the same time ("killing two birds with one stone") that part of the deficit attributed to recession, the cyclical deficit. The incorrect mix of the Memorandum – to wit, the constraining measures targeting the overall deficit – reduces the structural deficit but at the same time increases the cyclical deficit due to the recession it brings about. The distinction between cyclical and structural deficits is of crucial importance and is broadly adopted in applied economic policy, where it is used as a basic tool for evaluating the direction of a country's fiscal policy; for instance, whether it is expansive or contractive. The distinction allows us to assess the extent to which the attempted fiscal consolidation intensifies the problem of deficit, rendering the austerity measures ineffective. The critical part of Greek consolidation is the efficiency of fiscal measures in a period of deep recession, that is, the extent to which such measures yield results.

A recent study on debt sustainability by the Centre of Planning and Economic Research (KEPE, Athens, Greece) (May 2010), using the Hodrick-Prescott filter, attempts to divide the primary deficit for 2009 (i.e., without the interest) into a cyclical deficit of 3.9% of GDP and a structural deficit of 4.7% of GDP. The assessments of the economic team of ND are analogous, based on output gap methodology. Thus, the immediate implementation of compensating measures of low or nil budgetary cost and realistic assumptions[4] about the expenditure multiplier, the cyclical elasticity of the budget, Okun's law and so on will result in a quick recovery of the economy, reduction of unemployment and gradual elimination of the cyclical deficit. In technical terms derived from the business cycle theory, compensating measures are the "driving force" that will "pull" the economy out of the swamp. What are these measures? We mention just a few here (the complete list can be found in the "23 measures" proposed by ND): nil fiscal cost measures include concession contracts with self-financing (for ports, regional airports, road axes), in PPPs (Public–Private Partnerships)

[4] These assumptions are the elasticity of budget ranges between 0.6 and 0.7, the value of medium-to-short term expenditure multiplier ranges between 1.3 and 1.5, etc.

where the state participates with real estate, cabotage abolition and so on. Low cost fiscal measures include, among other things, enhancing residential activity, a particularly stagnant sector; at a fiscal cost of approximately €100 million (during the first critical two-year period), by subsidising house loans at 2%, GDP may be increased by 1%.

Moreover, it should be noted that "breaking" the vicious circle described above and converting the unfavourable psychological climate into positive expectations in the private sector can only be achieved through a quick recovery of the economy, which, through the so-called confidence multiplier, will result in second-round changes in expectations and income. The operation of the confidence multiplier is analogous to that of the expenditure multiplier. Therefore, just as the initial increase in government expenditure is translated into income for some, which through the marginal propensity to consume is converted in the second round into consumer expenditure and thereafter converted into income and so on, in this way the increase in national income in the initial phase of recovery will lead to positive changes in confidence, secondarily in an increase of income, leading to further positive changes and so on.

Finally, an important aspect of the alternative Economic Proposal is the enhancement of market liquidity. First, it is imperative that this peculiar suspension of payments from the state to the private sector be terminated and that the government immediately proceeds to pay its debts amounting to €7 billion. Second, the above-mentioned intermediate target should be complemented with a second intermediate target, that of establishing credit targets which will act as a guide for credit expansion.[5] Acknowledging the need to find a new, lower equilibrium point for credit, such a target could be a growth rate of 5% (over €10 billion).

Tackling the Structural Deficit

Unlike the Memorandum, the ND's Proposal has two basic premises for reducing the structural deficit: (a) the problem of ineffective performance measures cannot be solved by adopting more measures, since a structural problem calls for a structural solution to correct the problem at its root – for

[5] It is useful to remember that credit asphyxiation was the main explanation suggested by the current chairman of the Federal Reserve, Ben S. Bernanke (1983), for the long duration of the 1929 Great Depression.

instance tax evasion, which is the most characteristic cause of the structural deficit in Greece, cannot be reduced by increasing VAT rates but by providing incentives to pay VAT and establishing better tax control mechanisms; (b) fiscal consolidation should focus on the expenditure side (e.g., reducing public waste) and less on the revenue side, which is compatible with empirical evidences, such as examples of successful consolidation in other countries, but this is not the case in the Greek Memorandum. Thus, with the austerity measures already in place for 2010, amounting to approximately 9% of GDP, and with half of those included in the 2011 budget, provided that they will be effectively implemented, the target of gradual elimination of the structural deficit will be more than covered.

A direct consequence of this is that many austerity measures included in the 2011 budget – such as reductions in the solidarity allowance (fiscal result of €400 million), reduction in unemployment allowances (€500 million), freezing the pension indexation scheme (€100 million), cutting back social programmes (such as food aid, senior citizen care) and additional constraining of the PIB – will simply be unnecessary. Such drastic reductions are both socially unfair and contrary to growth. Taking into consideration that these measures are not in conflict with fiscal sustainability, while they are by nature anti-cyclical and with a high multiplier, there are strong reasons given the current circumstance of a deep recession – from the point of view of both social justice and economic efficiency – to cancel them entirely.

Tackling an Explosive Public Debt

Finally, unlike the Memorandum, which literally leaves the dynamic of public debt to its own fate, ND's Proposal acknowledges that the target of fiscal sustainability cannot be achieved by reducing the deficit alone, but that it is also necessary to drastically reduce public debt as a ratio of GDP, for many reasons, the most significant of them having to do with servicing the debt. In 2014, with a public debt in excess of €350 billion, we will be paying interest of approximately €20 billion per year, or 8.4% of GDP. With 80% of the debt being held by foreign entities, this means that 7% of GDP will flow abroad each year.

How can we ensure fiscal sustainability, namely, a decreasing ratio of debt to GDP? First, through a quick recovery of the economy and by achieving high growth rates (affecting the denominator); and second, through a generous one-off reduction of the absolute amount of debt through the resources derived from an optimal use and development of public property (affecting the numerator). A crucial parameter of ND's Economic Proposal is the exploitation of the other side of the balance sheet, notably the country's national assets, which would demonstrate that Greece is a "rich indebted" country. By establishing a General Secretariat of Public Property to supervise and coordinate such development in each of the three sectors of public property – chattels, real estate and intangible assets – it would become possible to raise funds in the near future in order to reduce public debt by 15–20% of GDP. We mention just some examples here: the portfolio of the state sector includes many corporations (*societés anonymes*) which could quickly generate many billions of euros through a bold programme of privitisation. Moreover, the concession of rights in the field of intangible assets (e.g., the gaming market and the digital signal frequency spectrum) could bring huge profits to the state. Finally, the development of public real estate, which is mostly inactive, could bring huge profits to the state and multiple benefits to the country's regional growth in general. The value of public real estate is evaluated at more than €270 billion and yet its internal rate of return for all these years has been almost nil!

Concluding Remarks

In conclusion, the immediate implementation of the threefold mix of ND's Economic Proposal with the coordinated and realistic policies it suggests would lead to a disengagement from the Memorandum, since this would have positive effects on the rating of Greek bonds by international agencies. It would also contribute to drastically reducing spreads and convincing markets that all of the preconditions for fiscal sustainability have been met, a fact that would enable us to borrow again at reasonable interest rates, which remains the desirable end result.

References

Bernanke, B. S. (1983). Nonmonetary effects of the financial crisis in the propagation of the Great Depression. *American Economic Review*, *73*(3), 1983, 257–76.

The Crisis in the Eurozone: Problems and Solutions*

Leszek Balcerowicz

The dramatic fiscal developments in Greece have provoked a huge and ongoing debate on how to deal with Greece's predicaments and the Eurozone problems. The discussions – and some of the decisions – have dealt with three overlapping problems:

- the EU's actual crisis management in response to Greece's debt distress;
- debt resolution mechanisms in the Eurozone; and
- long-term solutions for the Eurozone or, more broadly, for the EU.[1]

Let me start with the issue of "contagion". Some of the countries outside the Eurozone – Britain, Hungary, the Baltics – have seen a very serious worsening of their public finances. This, however, has not been widely perceived as an EU problem. In contrast, the dramatic fiscal developments in Greece have been widely described as a Eurozone problem. A reason for this difference appears to be the perceived danger of contagion in the Eurozone which would erupt if Greece defaults. There has been no such mention of contagion in the rest the EU in the debate on Britain's or Hungary's fiscal problems. It is evidently assumed that if Greece defaults the negative

---

* This text is derived from a lecture Professor Leszek Balcerowicz gave at an event organised by the Konstantinos Karamanlis Institute for Democracy in Athens, on 23 September 2010.

[1] I will leave aside here the proposals for how to deal with Greece's problem of insufficient external price competitiveness – the result of years of excessive growth in consumption, both public and private. These proposals envisage Greece temporarily leaving the Eurozone and then re-entering it at a more competitive exchange rate (Feldstein, 2010), the temporary introduction of a dual currency in Greece (Goodhart and Tsomocos, 2010) or radical reform which would lower labour taxes at the expense of increasing the VAT, etc.

spillover to other members of the Eurozone in the form of negative reactions on the financial markets could be especially serious (more orderly debt restructuring was not considered). What is the basis for such a claim? Is it the fact that these countries share the same currency, or the fact that some other members of the Eurozone, especially the large ones (Spain, Italy), are regarded by some as being in a bad fiscal situation (a large budget deficit in the former, a huge public debt in the latter)? If the first factors were the main reason, the fiscal problems of Ecuador (dollarised) or of Rhode Island (a small state in the US) would be a threat to a dollar. But such an assertion is absurd. It is not so much the strong linkages created by the monetary union per se, but the fact that some large members of the Eurozone may be perceived by the financial markets as fiscally vulnerable that is behind the fear of contagion. The solution to this problem is not only rapid fiscal consolidation in Greece but visible fiscal improvements in Spain and Italy.[2] This suggests that, while the Eurozone has uniform fiscal rules (see below), their enforcement should be especially stringent with respect to the larger members. However, the practice has been the opposite.

Among the EU members outside the Eurozone, Hungary and Latvia turned to the IMF for conditional crisis loans and obtained them. There was, to my knowledge, not much controversy about their action within EU decision-making bodies. In contrast, Greece's use of IMF crisis loans was the subject of heated debates in these institutions and was only accepted after much delay. It is hard to see the economic rationale for such double standards. If Hungary and Latvia were regarded to be in need of some conditional crisis loans, then this is no less true of Greece, too.[3] The EU and the Eurozone did not have at their disposal any ready-to-use institutional mechanisms for crisis loans which have the resources and the technical competence of the IMF (Pisani-Ferry and Sapir, 2010). One must ask whether the opposition to Greece's going to the IMF stemmed from the belief that no member of the Eurozone should ever use crisis loans as such or

---

[2] Such a consolidation, however, is first of all in Greece's interest. The lack of such an effort would matter for the Eurozone if it signalled to the financial markets that the same would happen in larger countries of the Eurozone (the signalling effect).

[3] It is worth noting that some EU countries which face a very serious economic and fiscal crisis, such as Estonia, Lithuania, Ireland, are coping with its consequences via tough economic adjustments and without IMF assistance.

that no Eurozone member should ever use IMF assistance. Both assumptions are difficult to defend on rational grounds. Therefore, one is left with the supposition that those who opposed the IMF option for Greece either erred in their economic reasoning or were guided by considerations of prestige which they do not clearly spell out.

After much delay and under the pressure of the financial markets, dramatic steps were taken on 7 May 2010 by EU decision-making bodies: the establishment of the European Stabilisation Mechanism (ESM) (discussed below) and the decision of the European Central Bank (ECB) to engage in outright purchases of government debt or, to be more specific, Greek debt. The latter step was widely – and I think rightly – perceived as a huge shock to the ECB's reputation. It was noted that "it is clearly inappropriate for any central bank to provide ongoing monetary financing for a sovereign which is no longer able to fund itself in the capital markets due to concerns about its solvency" (Mackie, 2010, p. 2). The ECB's explanation that buying Greek bonds was necessary to re-establish a "more normal" functioning of the monetary policy transmission mechanism is not very convincing, as it raises the question of what price of the bonds issued by the Eurozone countries is compatible with the "normal" operation of this mechanism. The future will tell whether and how the ECB will restore its credibility.

It is difficult to assess the consequences of the EU interventions, as the alternatives for dealing with the Greece's fiscal crisis were not spelled out and thus a necessary comparison was not made.[4] The alternatives were presented rather as an unmitigated catastrophe; this resembled the presentation of alternatives to bailing out large financial conglomerates or to a discretionary fiscal stimulus. However, even the proponents of the EU interventions admit that in themselves they did not provide a lasting solution but served rather to buy time. The question is, buying time for what?

Let us start with the debt resolution mechanisms. There are two main conclusions:

- There exist pure market solutions as well as the modified market solutions which are available for euro area countries. I don't see any reason why these countries should be banned from using them.

---

[4] Some authors (e.g., Subramanian, 2010; Dizard, 2010; Gros, 2010) argue that a better option would be the restructuring of Greece's debt, and that the EU interventions have only delayed this necessary step.

- The proposed Sovereign Debt Restructuring Mechanism (SDRM) does not offer any comparative advantage over other mechanisms (at least at the general level of analysis), and it is politically very difficult to introduce. It is hard to think of any Eurozone specificity which would justify its introduction in this area.

What about the Eurozone-specific crisis loan facility in the form of the ESM? It consists of a €60-billion rapid reaction stabilisation fund, controlled by the European Commission, and "a Special Purpose Vehicle", created by an intergovernmental agreement among Eurozone members, which will raise up to €440 billion on the market. The former component is guaranteed by the EU budget, that is, ultimately by EU members, the latter by the participating countries in proportion to the national share contributed to the ECB's capital.

One should not disregard the fact that the creation and use of the ESM required a rather heroic interpretation of Article 122.2 of the European Treaties, "which requires there to be exceptional occurrences beyond a member state's control for aid to be justified" (Buiter, 2010, p. 6). This step might contribute to the perception that the EU, while praising the rule of law, is not in fact a rule-of-law community, because it violates its own treaties.

However, it will be the operation of the ESM which will ultimately decide how it will be assessed. At the moment I can only ask some questions. First, the ESM's lending is to be conditional on a borrower-country's promised programme. What will be the relationship between this conditionality and that of IMF's? And a more general question: what is the intended relationship between the activities of these two institutions?

Second, as the ultimate fiscal responsibility for the operation of the ESM falls on the shoulders of the taxpayers in the participating countries, the final judgment of the ultimate authority, that of the citizens, especially in net-payer countries, would depend on whether they actually have to finance the losses of the ESM, that is, whether its conditional lending will turn into subsidising the less-disciplined Eurozone countries. This depends on the quality of the ESM's operation and how it affects the policies of the recipient countries. In discussing these issues one can draw on a huge literature on the IMF (see footnote 9).

Finally, let me turn to the issue of long-term solutions for the Eurozone. The creation of the Eurozone is widely thought to be an experiment, as it is a monetary union without political union, distinct from, it is claimed, previous monetary unions which were combined with political unions. This assertion is not very convincing. The countries which adopted the gold standard in the second half of the nineteenth century created a monetary union, in a broad sense, without a political union. To some extent the Bretton Woods system constituted a monetary union in the sense that its fixed-peg principle sharply limited the room for an independent monetary policy for its members. However, they clearly did not form a political union. Therefore, instead of focusing on "political union" as the requirement for a well-functioning monetary union, it is better to ask what is the broader set of conditions which determine the performance of any international monetary system based on hard pegs between the member countries' currencies.

First, these monetary systems all required fiscal discipline from their members. This was the case under the gold standard, with its informal norm of a balanced budget, until both the norm and the gold standard unravelled under the shocks of the First World War and then the Great Depression. Both the norm of a balanced budget and the gold standard were later de-legitimised among the elites because of the expansion of Keynesianism.

Second, the monetary unions consisting of sovereign states existed without any fiscal transfers from a common centre, because such a centre did not exist. Instead, besides the norm of fiscal discipline, the economies of the members of successful monetary unions of this kind displayed a great deal of flexibility, including – and this is especially relevant – in labour markets. This was important because it facilitated and shortened adjustments to asymmetric shocks.

The meaning of the term "political union" in the debate on the Eurozone is often vague. My view is that in the context of the debate on the monetary union the political union should be defined as having at least two components:

First, the members of the union have limited fiscal sovereignty; that is, there are some institutional limits on their deficits and/or debt. Within a sovereign country, which is the strongest version of a political union, regional and local governments do not have complete freedom in this respect.

Second, there is a substantial common budget so that those parts of the political union which are hit by asymmetric shocks would get some transfers from this budget via automatic fiscal stabilisers or discretionary spending.

The first component constitutes the preventive arm of the political union; that is, it aims at forestalling fiscal threats to the value of the common currency.[5] The second component is the protective arm of the political union. That is, it is designed to protect the population of the most-affected regions from deep declines in consumption. I have the impression that those who claim that "political union" is necessary (or at least desirable) for a monetary union have mostly in mind the second component and disregard the first, despite the fact that fiscal constraints on local governments are clearly a typical and important component of single sovereign states, the strongest form of political union. In this sense they ignore the fact that the Stability and Growth Pact has been, in principle, an important component of the political union and not its substitute (for the importance of this Pact see Tanzi, 2004).

Against this background, it is useful to look at the creation and evolution of the Eurozone. To cut a long story short: fiscal criteria (the Stability and Growth Pact) were rightly introduced, as they constitute a preventive arm of the political union important for the monetary union. In addition, at the insistence of Germany the bail-out clause was introduced in order, I think, to strengthen incentives for fiscal discipline in the respective members of the Eurozone and to avoid Argentinian-type developments, whereby the fiscal irresponsibility of the provinces undermined the fiscal and monetary stance of the whole union. However, from the very beginning these safeguards have been eroded by the largest countries of the Eurozone (Germany, France) which obviously have had the crucial say in the decisions regarding the creation and evolution of the Eurozone.

First, the original sin was committed: countries which were in violation of the fiscal criteria (Italy, Belgium and probably Greece) were admitted into the Eurozone. Second, Germany and France breached the Stability and Growth Pact. Third, in response to that breach, the Pact itself was modified;

---

[5] Argentina provides a warning against a political union where the regions are not constrained in their fiscal policy (see Besfamille and Sanguinetti, 2003). However, one should add that even in the US some states run into huge fiscal problems; see, for example, California during the present recession in the US.

in other words, it was made more flexible. Many economists criticised what they perceived to be the macroeconomic imperfections of the Pact and welcomed its modification. I think these economists missed the essential point, however: rules which are violated and then quickly modified by the largest members of the club cease to be rules at all, that is, they cease to be the binding constraints for the members of the club. I don't want to say that external pressure is sufficient to make members respect the agreed norms of fiscal discipline. On the contrary, there is no good substitute for domestic frameworks for the fiscal behaviour of governments and – what ultimately matters – for the strong representation of fiscally conservative voters in the respective countries. However, the violation and then modification of the agreed common norms of fiscal conduct were a bad example and thus made the emergence of such domestic frameworks and of the appropriate structure of public opinion much more difficult.

The final stage in our short story is the present global financial crisis, which has revealed and deepened in the Eurozone the consequences of previous vulnerabilities: highly expansionary fiscal policies (especially in Greece) and housing booms (Spain, Ireland).

However, simply deploring past errors is not the proper way to deal with their consequences. What is needed instead is to learn from these errors. It should be amply clear that there is no scope in the foreseeable future for the extension of the EU budget so as to strengthen the protective arm of the Union via increased fiscal transfers to members affected by deep declines in consumption. The increased role of such a budget requires an enhanced level of group identity which cannot be artificially generated by the political elites. And the EU, given the separate histories of its member states, is certainly a long way from a strong European identity among its respective societies. What is more, as rightly stressed by Otmar Issing (2010), any attempt by the elites to engineer bail-outs of members which are in clear breach of the commonly agreed rules would provoke a storm in at least some countries, thus depressing instead of enhancing the level of "European solidarity".[6]

---

[6] Even within individual countries, large transfers from one part to another which are perceived to pay for inefficiencies and waste are likely to produce social and political tensions, as shown by those between northern and southern Italy and, perhaps, by West and East Germany.

However, the crucial point is not that it is politically impossible to create an enhanced protective arm within the Eurozone; the key point is that this would not address the main problem, which is the weakness of its preventive arm and, more broadly, of mechanisms safeguarding the fiscal discipline in respective member states.

Instead of looking at the wrong model, that of a single state, EU institutions and countries should focus on what are the conditions for a proper functioning of the right model, namely a gold standard–type of monetary union, a union of countries with a single currency but without any larger common budget to compensate for asymmetric shocks. While doing so one must consider, of course, some later developments which are or should be present to strengthen these conditions.

These conditions can be grouped into three categories:

1. Mechanisms to prevent procyclical policies and severe fiscal shocks. These mechanisms should operate both at the level of the EU (and the Eurozone) and at the level of the respective countries.
2. Structural reforms which would strengthen their long-term growth. These are necessary not only for the continued improvement in the standard of living of the populations but also to help them to grow out of increased public debt (see White, 2010).
3. Structural reforms to facilitate the adjustment of the economy to various shocks.

In the first category the following measures appear to me to be most important:

- Accounting rules, which define budgetary deficits and public debt, must be made credible and transparent. Enron-type accounting should have no place among either companies or governments. The rules should consider not only the explicit debt but also the implicit debt (e.g., pension liabilities).
- Monitoring of budget deficits and the public debt must be strengthened. This is a job for Eurostat, the European Commission and the European Risk Council proposed by de Larosière report (2009). Monitoring should also focus on the development of asset bubbles which, when they burst, can produce deep recessions and the resulting sharp increases in budgetary deficits.
- The Stability and Growth Pact should be enforced, which implies the use of available sanctions. These should be strengthened, if possible.

- The monetary policy of the ECB should pay more attention to the developments of asset bubbles, which, when they burst, can produce huge fiscal shocks. In other words it should be a more conservative policy than one which is guided only by inflation as measured by the Consumer Price Index (CPI). It could mean that the ECB makes more use of its monetary pillar in its decisions on interest rates.[7]
- The ECB's common monetary policy cannot fit the macroeconomic conditions of all the member countries. For example, the ECB's interest rates were too low for Spain or Ireland, which contributed to the development of asset bubbles in these economies with the resulting bust, recession and large increase in their public debt. Therefore, Eurozone countries (and other countries, too) need an additional instrument, macroprudential regulations which aim at reducing the excessive growth of credit. While the need for such regulation is now widely recognised, much technical work remains to be done.
- The initiatives at the EU and/or the Eurozone level cannot substitute for the strengthening of preventive mechanisms in the respective countries. This is ultimately the responsibility of domestic politicians and the public at large. However, disciplinary measures at the EU level are desirable and, perhaps, even necessary to spur the growth of preventive mechanisms in the respective countries. As EU initiatives are largely dependent on large countries, they bear a special responsibility for developments in the Eurozone – and in the EU.

The second category would include the following main steps:

- At the EU level, probably the most important mechanism for the long-term growth of all member states is the single market. Economic nationalism which risks damaging it must, therefore, be prevented at all costs. This is necessary but not sufficient. The vigorous effort to complete the single market should be relaunched. This applies first of all to non-financial services, where there is the largest gap vis-à-vis the US.
- The Lisbon Agenda should focus on economic goals and be reinvigorated. This should mean more market reforms, not setting numerical tar-

[7] The best solution would be that the US Federal Reserve, another globally important central bank, change its approach too. Otherwise, the more conservative ECB policy would lead to the appreciation of euro.

gets that make little sense; for example, requiring that all countries spend 3% of their GDP on research and development.[8]

- EU institutions and countries should reconsider measures which risk imposing additional burdens on their economies and/or hamper the flexibility of markets. I have in mind, first of all, the EU's climate policy, which has a weak analytical basis and has been presented as though it were offering a free lunch. The drift towards social policy increasingly becoming the EU's responsibility should be stopped, as it risks introducing additional rigidities and burdens in the more flexible economies and raises fundamental constitutional questions (the subsidiarity principle).[9]
- Fiscal reforms in the respective EU countries are not only fundamentally important in the short run – that is, to deal with increased budgetary deficits and public debts – but in the longer run too. Persistent deficits and a large public debt are detrimental to long-term growth, because sooner or later they crowd out private investment and introduce harmful uncertainty, which worsens the investment climate. The mode of fiscal consolidation also affects the forces of growth: as all EU members already impose a large tax burden, further tax increases would weaken those forces. The focus of fiscal reform should, therefore, be on measures which reduce the growth of spending commitments, which – given the ageing of the EU's societies – must include pension reforms that raise the age of retirement.

Finally, regarding the third category let me note that some research suggests that fiscal adjustments in the absence of the depreciation option, that is, under hard pegs, may be more difficult than those when the depreciation or devaluation option exists (see Lambertini and Tavares, 2005; Mati and Thornton, 2008). However, this is obviously not a reason to scrap the EMU but an argument for measures to prevent serious fiscal imbalances in the first place (category 1). It also increases the importance of structural reforms which would facilitate the adjustment of the economy to various shocks, including fiscal consolidation. Rigid (or dual) labour markets and, more generally, rigid prices and regulatory constraints on the

---

[8] This target disregards the differences in the level of development: economies with larger per capita income can use more technology transfer and thus need to spend less on R&D. Besides, increased R&D should result from reforms which increase the scope of markets and intensity of competition.

[9] For more, see Klecha (2008) and Threlfall (2007).

supply response of the economy deepen its recessionary reaction to various shocks and contribute to the growth of unemployment. Therefore, liberalising reforms should be a priority wherever needed and should be the other focus of the reinvigorated Lisbon Agenda.

No amount of exclamations about the "European solidarity", "social cohesion" or the "European social model" can substitute for these reforms, especially because, as I have already stressed, an EU-wide protective fiscal policy is not in prospect, and even if it were it would not provide the proper response to the main problem: the weakness of the preventive mechanisms in the EU.

References

Besfamille, M., and Sanguinetti, P. (2003). Formal and real fiscal federalism in Argentina. *CSGR Working Paper No. 115/03*, May. The University of Warwick.

Buiter, W. (2010). Can the European stabilization mechanism save the Monetary Union? *City Economics*, 20 May.

de Larosière, J. (2009). *The High-Level Group on Financial Supervision in the EU*. Brussels.

Dizard, J. (2010). A Greek restructuring to avoid Greek tragedy. *Financial Times*, 10 May, 10.

Feldstein, M. (2010). Let Greece take a holiday from the Eurozone. *Financial Times*, 16 February, 9.

Goodhart, C., and Tsomocos, D. (2010). The Californian solution for the Club Med. *Financial Times*, 25 January, 9.

Gros, D. (2010). The euro can survive a Greek default. *The Wall Street Journal*, 26 April, 13.

Issing, O. (2010). A Greek bail-out would be a disaster for Europe. *Financial Times*, 17 February, 9.

Klecha, S. (2008). Europe's social policy as a difficult negotiation process. *IPG*, *1*, 68–87.

Lambertini, L., and Tavares, J. A. (2005). Exchange rates and fiscal adjustment: Evidence from the OECD and implications for the EMU. *Contributions to Macroeconomics*, *1*, 1–28.

Mackie, D. (2010). Huge policy intervention in the euro area. *J. P. Morgan, Europe Economic Research*, 10 May.

Mati, A., and Thornton, J. (2008). The exchange rate and fiscal consolidation episodes in emerging market economies. *Economic Letters 100*, 115–18.

Pisani-Ferry, J., and Sapir, A. (2010). The best course for Greece is to call in the fund. *Financial Times*, 2 February, 9.

Subramanian, A. (2010). Greek deal lets banks profit from "immoral" hazard. *Financial Times*, 7 May, 9.

Tanzi, V. (2004). The Stability and Growth Pact: Its role and future. *Cato Journal*, *24*(1–2), 67–9.

Threlfall, M. (2007). The social dimension of the European Union: Innovative methods for advancing integration. *Global Social Policy*, *7*, 271–93.

White, W. (2010). We need Plan B to curb the debt headwinds. *Financial Times*, 3 March, 9.

Fiscal Rules in the EU: Time to Rethink and Start from the Basics

Pyrros Papadimitriou

Introduction

The economic literature provides a wide range of analysis on how fiscal aggregates in general and efficiency in the public sector in particular affect the economy. However, economic history has shown that fiscal and public policies often depart from what could be considered optimal. The large increases in debt ratios over recent decades and the deficit bias of some governments are well-known examples of such sub-optimal behaviour. The causes of the deficit bias have been discussed extensively in the literature[1] and appear to include political economy considerations related to policy-makers' short-term thinking and voters' fiscal illusion.

Today there is a broad consensus that the domestic institutional setting of a country – that is, the procedural rules governing the budget process, the numerical fiscal rules and independent institutions – are of great importance for the conduct of fiscal policies. They create the environment, the incentives and the constraints under which fiscal policy decisions are taken. Moreover, more emphasis is being given to the "microfoundations" of fiscal policy as a way to promote the efficiency of any single payment made by the state.

Such arrangements are of particular importance in the context of the European Economic and Monetary Union (EMU) due to the adverse spillover effects that undesirable fiscal policies may have. In September 2010, under the pressure of the fiscal difficulties in Greece, Ireland, Portugal and Spain, the European Commission presented a package of proposals aimed at reforming economic governance in Europe. However, member states

---

1 For a review see Krogstrup and Wyplosz (2009).

seem reluctant to accept them and doubts about the long-term prospects of the euro have increased.

The purpose of this paper is to show that as long as fiscal policy remains within the purview of national authorities and until a new system of fiscal governance is established some basic fiscal rules and procedures could have a significant role in improving fiscal situation in the Eurozone. So far, the Stability and Growth Pact (SGP) has proved to be ineffective in restraining member states from running excessive deficits, mainly because it cannot alter the incentives for governments.

In the first section of this paper, I review the experience of some OECD countries which have shown that there are other more appropriate methods for changing governments' incentives than strict surveillance and penalties. In the second section we present the Greek experience in implementing budgets in the period 2002–8. The substantial deviation in actual payments from those recorded in the budget reveals that budgets in Greece could be characterised as "indicative", thus showing a lack of budgetary procedures in the Eurozone. These procedures should have been applied much earlier or at least simultaneously with the SGP. In the third section I examine the impact of the SGP to change governments' incentives. Indeed, the language of "preventive" and "corrective" mechanisms indicates that the intention of the Pact was to affect the incentives that governments face when they make fiscal policy decisions. However, experience has betrayed these intentions. In the fourth section the emphasis is on a series of measures that should be regarded as a prerequisite for any reform in the fiscal governance of the EU. Although these measures are not as ambitious as the ones presented by the Commission, they can easily be implemented without running the risk of not being implemented, as was the case with the SGP. Finally, section five summarises the results.

Fiscal Rules, Performance Management and Public Reporting: The Experience of OECD Countries

In OECD countries there is a wide range of arrangements aimed at improving fiscal performance. Rules for the budgetary process, numerical fiscal targets, independent public bodies, performance management indices and effective reporting to the public are some of the prevailing practices. The chosen practices, to a large extent, depend on the history and the institutional characteristics of each country and on the nature of the fiscal problems.

The Budgetary Process

The budgetary process is described as a number of procedural rules governing the drafting of the budget, its passage through parliament and its implementation. These rules distribute strategic influence among the participants in the budget process and regulate the flow of information. A key element is the distribution of powers between the government and the legislative branch, that is, which has the power, and under what conditions, to amend the budget and the constraints on discretionary policymaking.

In the literature there is sufficient evidence that sound procedural rules help fiscal performance (von Hagen, 1992; von Hagen and Harden, 1995; Schiavo-Campo and Tommasi, 1999; Poterba and Rueben, 2001; Gleich, 2003; Ylaoutinen, 2004). Moreover, there is evidence that fiscal discipline is enhanced by budget procedures in which the finance minister is strongly dominant over spending ministers and where the flexibility in the execution and the amendment power of the parliament is limited (von Hagen, 1992; Baldwin, Gros and Laeven, 2010).

Numerical Fiscal Rules and Targets

Fiscal policy rules have gained prominence in Europe and elsewhere in the past 20 years. Such rules specify numerical targets for key budgetary parameters. Limits imposed on the deficit or debt of governments and constraints imposed on some categories of public expenditure or tax revenue can be considered as fiscal rules.

However, it is difficult to set an ideal rule. According to Kopits and Symanski (1998) the ideal rule is well defined, transparent, simple, sufficiently flexible, adequate to the final goal, enforceable, consistent and underpinned by structural reforms. In this context it is obvious that trade-offs between these characteristics are prevailing, for example, between enforceability and flexibility or between simplicity and adequacy to final goals. Moreover, under fiscal rules it is difficult for economic policy to deal with unexpected circumstances and changes in the economic situation (Wyplosz, 2002). The difficulty is then to find the best balance, taking into account the country-specific environment.

Independent Institutions

The third type of institutional reform to address the deficit bias consists of independent, non-elected public bodies which complement the existing national institutions. These institutions are designed to ensure an appropriate use of discretionary power in the conduct of fiscal policies. The main issue, however, is that these judgements are not in the hands of elected politicians.

The literature proposes a number of criteria to evaluate whether some degree of economic policy delegation to independent institutions would be desirable (Alesina and Tabellini, 1990). First, political considerations (e.g., electoral cycles) could lead to harmful distortions in policymaking. Second, there must be a broad and stable consensus on what sound policy consists of. Third, the delegated mandate must not have any distributive consequences, since distributional decisions can be exercised only by elected representatives. Fourth, delegation should not lead to conflict with other policy areas. While these four criteria are broadly met in the case of monetary policy, this is less obvious for fiscal policy. This is because fiscal policy has to deal with a complex trade-off between sustainability and stabilisation, and because almost all fiscal policy decisions have redistributive consequences.

The inability to (a) reach solid conclusions about the role of institutions in improving policy outcomes;[2] and (b) reach a consensus on whether there is a case for delegating part of fiscal policy to non-elected independent bodies has led to more modest proposals. These proposals, which have already been accepted by many OECD countries, recommend the establishment of institutions that can ensure that fiscal policy is based on accurate data, provide analysis on fiscal policy issues and release regular assessments and recommendations, notably with a view to increasing the "reputation costs" of conducting unsound policies.

---

[2] See Schick (2004) for a review of the literature.

Performance Management

Performance management has never been more critical for the public sector than it is today. The past two decades have witnessed a growing interest in performance management and budgeting reforms in response to louder public demands for government accountability. These reforms are intended to transform public budgeting systems from the control of inputs to a focus on outputs or outcomes, in the interest of improving operational efficiency and promoting results-oriented accountability.

To facilitate this process many countries have introduced legislation and frameworks to improve the performance of their government. In the US, for example, each public agency is obligated to submit to the Office of Management and to Congress a strategic plan with specific performance indicators. The key performance results are then aggregated into an executive scorecard, which is easily accessible to the public.

Similar initiatives have been undertaken by the government of the UK (Best Value Performance Indicators, and more recently the Comprehensive Area Assessment [Audit Commission 2010]), Canada (client satisfaction) (Treasury Board of Canada Secretariat 2010), Australia (customer service charters) China, Sweden and the Netherlands. What most of these initiatives have in common is that they provide frameworks for measuring performance and have predetermined performance indicators. Moreover, many governments use league tables or performance scorecards to inform the public effectively.

Public Reporting

Complementary to the policies for improving the performance of the public sector are policies that aim to improve public reporting. Public reporting was developed by public administration theorists in the first half of the twentieth century in an effort to bring the emerging administrative state into harmony with democracy.[3] Public reporting is defined "as the management activity intended to convey systematically and regularly information about government operations, in order to promote an informed citizenry in a democracy and accountability to public opinion. It consists

---

[3] For a review of the literature see Lee (2005).

of direct and indirect reporting of the government's record of accomplishments and stewardship of the taxpayers' money" (Lee, 2004, p. 5).

In recent years, the combination of digital technology with a focus on performance measurement has provided an opportunity to reformulate the original conception of public reporting. For many researchers, so-called e-reporting is one of the ways to restore public trust in government (Herzlinger, 1996). The OECD issued two reports in 2001 calling for improved public reporting and providing guidelines for such efforts (Caddy and Vergez, 2001; Gramberger, 2001).

Implementation of the Budget: The Greek Experience in the Period 2002–8

In Greece the budget has never been perceived as the "absolute" policy tool it usually is in other countries. The whole discussion about the annual budget is limited to the voting procedure in Parliament. The methodology of elaborating the budget as well as its implementation in the previous year are inadequately discussed or not discussed at all. Most Greek budgets of the past were in fact not binding and merely a simple depiction of expected revenues and expenses. As a result one could easily characterise them as "indicative". In many cases, the level of expenditures shown in the following paragraphs has been determined ex post facto. New expenditures that were not part of the budget were undertaken by simple government decision.

Table 1 shows the deviations in actual expenditures from those recorded in the budget, according to the Greek Ministry. To estimate the deviations, data have been collected from the annual budgets and the Budget Execution Bulletins (General Department of Treasury and Budget). Excess expenditures are depicted by a positive sign; when actual expenditures are less than estimated in the budget, the sign is negative. The estimates reveal substantial deviations in actual expenditures compared to the forecasts. The case of the Ministry of Transport and Networks is indicative. For the whole period 2002–7 actual expenditures exceeded forecasts substantially.

Table 1. Budgeted versus accrued expenditures (%)

Ministries – Functions	2002	2003	2004	2005	2006	2007	2008
Presidency of the Hellenic Republic	-3	-7	+5	-4	-2	-3	-4.9
Parliament	+3	0	+8	-1	0	+2	-3
Ministry of Interior and Decentralisation	+16	+12	+11	+1	+2	+2	-1.5
Ministry of Foreign Affairs	+23	+6	-8	+4	-3	-11	-13.9
Ministry of National Defence	+8	+9	+4	-5	+9	+2	-0.2
Ministry of Justice	0	0	0	-2	-4	-4	-1.9
Ministry of National Education and Religious Affairs	+2	+3	+10	+3	-1	0	+1,6
Ministry of Health	+4	+8	+5	+59	-1	+3	+1.4
Ministry of Culture	+8	+6	+8	+4	+13	+14	+3.4
Ministry of Finance, excluding General Public Expenditures	-3	+18	+19	+7	+13	-	+6.2
Ministry of Finance, General Public Expenditures	+1	+1	+11	-3	-2	-4	-3.9
Ministry of Macedonia-Thrace	-1	+19	-2	-10	-8	-7	-12.8
Ministry of Environment, Energy and Climate Change	+8	+6	+12	+2	-3	-3	+3.5
Ministry of Transport and Communications	+31	+31	+53	+32	+25	+27	-0.5
Ministry of Merchant Shipping	+1	+10	+25	+5	+3	+7	-
Minister of the Press and the Media	+5	+19	+62	+18	+13	+13	-4.5
Ministry of Employment and Social Security	-3	+2	+12	+5	+6	-3	-1.6
Ministry of Development	+5	+12	+16	+16	+18	+45	+3.1
Ministry of Public Order	+3	+3	+21	-1	-3	+2	-

Greater deviations are found if we compare the data at a higher level of analysis (e.g., four digits). From Table 2 it is clear that the salaries of temporary personnel and the subsidies to mental health units were systematically underestimated. Moreover, the distinction between estimates in the budget, the actual obligations of the State and expenditures should be emphasised. In Table 2 the significant difference between obligations and expenditures is obvious. The Greek government seems to postpone payments, exploiting the absence of accounting practices in the public sector similar to international accounting standards in the private sector.

Table 2. Deviations in the budget in selected budget lines (in euros)

	Salaries of staff employed under terms of private law (seasonal staff is included) (Code 0342)		
	Estimates	Obligations	Payments
2009	177,000.00	8,797,062.00	-
2008	214,100.00	12,559,912.00	10,752,449.70
2007	342,300.00	14,467,120.00	10,234,218.21
2006	249,500.00	8,463,449.00	7,566,314.69
	Social insurance payments for the above staff (Code 0352)		
	Estimations	Obligations	Payments
2009	63,000.00	2,421,288.00	-
2008	71,000.00	3,466,210.00	2,814,224.94
2007	103,800.00	4,083,132.00	2,645,484.28
2006	75,000.00	2,359,860.00	1,943,959.47
	Grants to mental health units (Code 2544)		
	Estimations	Obligations	Payments
2009	40,000,000.00	70,907,000.00	-
2008	38,000,000.00	45,800,000.00	47,564,794.53
2007	23,000,000.00	45,017,735.00	54,247,485.00

The observed deviations reveal that the rules governing the budget process in Greece are weak. Under these circumstances, and irrespective of the international financial crisis, it should be expected that Greece, sooner or later, would land in fiscal distress. However, the European Union itself also bears significant responsibility for the current fiscal situation in member countries.[4] The establishment of a monetary union without effective fiscal rules inevitably led to such outcomes. The insistence on general quantitative indices in the SGP, without any discrimination by subcategories of revenue and expenses and without procedures for monitoring them, had to lead sooner or later to macroeconomic distress.

[4] This applies independently of the fact that the causes of the deterioration of Greek public finances were different from those in Ireland and Spain.

The Stability and Growth Pact: Poor Performance and Wrong Incentives

In the EU, the Maastricht Treaty and the SGP (1997, and its reform in 2005) imposed well-known budgetary obligations on the member states. However, the track record of member states in respecting them varies considerably. The effect of the SGP on the sustainability of public finances has been less positive than commonly believed. In the first stage governments fell into the trap of overlooking the long-term effects while focusing on the short-term benefits. Eventually, due to the financial crisis, among other factors, the deficit of the Eurozone increased from 1.6% in 1999 to 6.3% in 2009.

The insufficiency of the SGP to create conditions of fiscal stability in the member countries has been attributed to various factors. A monetary union per se induces a bias towards fiscal laxity. This risk derives from (a) the theoretical argument that fiscal policy under a fixed exchange rate is more effective;[5] (b) the disappearance of the closed economy crowding out the effect of expansionary fiscal policies; and (c) the removal of the threat of open economy exchange-rate crises.

Even before the introduction of the euro, various countries used techniques of creative accounting to reduce the deficit reported to Eurostat, using "innovative" one-time transactions – which allowed spending without impacting the recorded deficit – such as securitisation, financial derivatives, one-time payments by state related entities and so on. The possibility that candidate countries in the pre-euro era could follow such policies had been pointed out at an early stage by various economists (Buiter, Corsetti and Roubini, 1993; Easterly, 1999). However the reality in the post-1999 period has exceeded expectations. As long as EMU membership was not secure, voters rewarded signs of fiscal discipline as this would increase the chance of getting into the monetary union. Governments had the incentive to undertake discretionary fiscal contractions even in election years. Once EMU membership was secured, the old pattern of political budget cycles re-emerged.

---

[5] There is debate concerning the expansionary effects of fiscal policy and especially when this involves permanent deficits. Moreover, it is often argued that fiscal consolidation has an expansionary impact on economic activity. For a review see Ferreiro, Fontana and Serrano (2008, p. 86).

Koen and van den Noord (2005) and von Hagen and Wolff (2004) provide evidence that one-time measures have been used more frequently since the inception of the EMU and proved that their probability is correlated with the magnitude of the deficit. As Coeuré and Pisani-Ferry (2005) point out, there have been outright disposals of public assets with the sole aim of lowering the gross debt. There have been more devious operations aimed at substituting on-balance debt for off-balance liabilities. Some countries have cashed in immediate revenue in exchange either for additional pension liabilities (France Telecom and EDF transfers in France, postal pensions securitisation in Germany) or for lower future revenues (Italian, Portuguese and Greek securitisations).

These practices raise questions concerning (a) of the rightness of the disposal price of public assets; and (b) of the net outcome for the wealth of the member states. According to Milesi-Ferretti and Moriyama (2004), who have investigated empirically the dynamics of EU governments' valuations, member states were poorer after the establishment of the SGP. They compared changes in financial and non-financial assets with changes in financial liabilities and corrected for valuation effects. They showed a sharp contrast between the periods 1992–7 and 1997–2002. In the first period, increases in public liabilities were matched by changes in assets and the net value of governments was relatively stable. This was not the case in the second period. EU governments were poorer in 2002 than in 1997.

Based on the short-term availability of funds to finance the deficits, many governments overlooked sustainability. It is known that in the long run, under the "intertemporal budget constraint", the discounted sum of a government's expected expenditures cannot exceed the discounted sum of its expected revenues. In other words, to achieve sustainability governments should make reference to the conditions of today's balance sheet and to future revenues or liabilities.[6] However, governments' incentives to follow such policies were restricted under the SGP. By putting emphasis on partial criteria such as deficit and debt, the Pact reinforced governments' myopia (Coeuré and Pisani-Ferry, 2005) and has added to the difficulty of structural reforms, at least those which imply short-term budgetary cost.

---

6 For an extended analysis see Buiter and Grafe (2002).

Under these circumstances one can seriously consider the case that the SGP did not help the Eurozone increase its long-term growth rates. By treating all expenditures the same way, the SGP suppressed incentives to carry out productive public investments (Blanchard and Giavazzi, 2004; Coeuré and Pisani-Ferry, 2005). The emphasis on maintaining budget positions "close to balance" implies that capital expenditure has to be funded from current revenues. Debt finance will no longer be available to smooth the burden of investment projects over the generations. The choice between tax and debt financing of government investment thus affects the distribution of welfare across generations. SGP provisions implied a disincentive to undertake projects producing deferred benefits, and obviously this disincentive is stronger during consolidation periods.

Last but not least, the SGP very early became dysfunctional when it was evident that the Council would not impose sanctions on countries with excessive deficits. Furthermore, even from the first years of its implementation, the governments of Germany, France and other countries started pushing for a reform of the SGP, asking for more "flexibility".[7] It is in this context that long discussions are now taking place for more fundamental revisions of the EU fiscal framework, including proposals to scrap the SGP altogether.

Reform Proposals: Starting from the Basics

On 29 September 2010, the European Commission presented proposals for the reform of economic governance in the EU. The package contains six legislative proposals, including a second reform of the SGP and macroeconomic coordination. These proposals aim to solve some of the above-mentioned problems of the SGP but they seem difficult to accept for the member countries.[8]

Experience so far indicates that the decision-making process in the EU is a quite complicated issue. In this context, measures such as the ones pre-

[7] Somewhat ironically, Germany, the very country that had pushed for tighter fiscal rules in the EMU in the mid-1990s, was the second EMU country to violate the fiscal rules.

[8] Italy opposes the new focus on accumulated debt; France dislikes the idea of "semi-automatic" sanctions; Spain rejects the notion of penalties against countries deemed to be losing competitiveness.

sented below could possibly contribute to providing the necessary conceptual and accounting infrastructure, until more ambitious amendments to fiscal governance in the EU take place. These proposals could easily be implemented in practice. They have the advantage of providing sufficient information to the public, thus creating the right incentives for governments to implement sound fiscal policies. As is commonly accepted, institutional reforms per se do not change policymakers' preferences and in a democracy the effectiveness of any arrangement depends on the degree of political support and the existence of a wide social consensus.

Unified Public Sector Accounting Rules and More Comprehensive Balance Sheets

The experience of many member states that used creative accounting techniques to reduce the deficit reported to Eurostat suggests the need for (a) all EU member states to adopt unified public sector accounting rules and standards; and (b) more comprehensive government balance sheets, which will not allow governments to turn on-balance into off-balance liabilities.

To achieve this goal it is important to establish a methodology that will help sustainability assessment and to select some variables for monitoring the fiscal situation of governments.

The net value of the government sector, that is, the difference between its total assets and its financial liabilities (excluding implicit liabilities), is a variable that helps in sustainability assessment.[9] This is the closest equivalent to a company's equity. Non-financial government assets are known to be difficult to define and value. They are frequently non-marketable and when they are, valuing them on the basis of future cash flows or of liquidation value makes quite a difference. However, the proper management of a government's balance sheet requires a fire sale of public assets. There is therefore a case for taking into account marketable assets at least (not a historic place in Athens but certainly public real estate).

Implicit liabilities such as pensions cannot be aggregated to financial liabilities because they belong to a different class of debt. As is properly pointed out by Eurostat (2004), unfunded pay as you go pension schemes cannot be treated as on-balance liabilities since "their value can be unilaterally altered by the debtor." Their present value can "jump" as a conse-

[9] For a more extended discussion see Coeuré and Pisani-Ferry (2005).

quence of parametric reforms or changes in growth assumptions. Therefore, pension liabilities should not be added to conventional debt but should be used to complement debt and deficit indicators (Oksanen, 2004; Coeuré and Pisani-Ferry, 2005).

The above methodology resembles the international accounting standards of the private sector. Several EU governments are publishing their assets and liabilities under international accounting standards, following pioneering countries outside the Eurozone (e.g., New Zealand, Australia, the US, the UK and Sweden). France also has since 2006 published an opening financial statement, which is slightly different from private accounting, but very clearly presents the French general government "equity", that is, its net value.

Provided that unified accounting rules are applied to all EU countries and that indicators of sustainability and the fiscal situation are properly estimated, the incentives of governments may change. Defining sustainability as the net value of the government as a percentage of GDP at a certain point in time will allow the average voter to give more attention to the fiscal implications of public policies.

Evaluating Each Payment: The Purpose of Each Payment and Its Opportunity Cost

The notion of opportunity cost has to be more prominent in fiscal policy. Today it is more than obvious that the positive impact from public expenditure comes from its composition, not from its size. Consequently, the composition of public expenditures can be targeted in order to influence the growth rate of economic activity. An in-depth and real opportunity cost analysis can help in the re-composition of public expenditures, increasing the share of "productive" expenditures, namely, expenditures with a multiplier impact on inputs (capital and labour) and productivity of these inputs.

Public policy endogenous growth models are the theoretical basis of this new fiscal policy strategy. These models focus on the role that fiscal policy can play in enhancing or retarding economic growth. Since in endogenous growth models economic growth is determined by inputs and technical progress, "productive" expenditures are those that by complementing private sector production and generating positive externalities to firms have a positive effect on the marginal productivity of capital and labour, and "unproductive" expenditures would be those that give direct utility to households.

Although the empirical evidence in support of endogenous growth through fiscal policy is mixed, a re-evaluation of all public expenses, for each budget line, can reduce extravagance. Moreover, if associated with e-reporting practices such as publishing on the web the purpose of each payment and its opportunity cost, one can easily expect that the economy will gain in fiscal efficiency. Furthermore, this can be enhanced if public authorities provide ongoing evaluation reports, namely, to what extent the pre-announced targets have been fulfilled. The same analysis of opportunity costs can be attached to any decision about holding government assets or privatising them.

Budgetary Procedures

In the second section, evidence was presented that sound budgetary procedural rules help fiscal performance. However, at the moment each member state has only the obligation to report, which means that it has to inform the Commission after the completion of its budget.

Under the present circumstances, each member country must develop a system of rules and procedures for its budget – if such a system does not yet exist – adjusted to a set of predetermined general conditions that will be accepted by all member countries. This system should govern the elaboration and the implementation of the budget, fix the respective powers of the various actors taking part in the budget process, determine the pre-conditions for amendments and allocate the power to amend. As far as its implementation is concerned, any amendments or deviations in the budget lines should be announced in advance to the European Commission.

Conclusions

This paper has shown that fiscal governance in the EMU is far from satisfactory. Apart from strict surveillance and penalties, there are definitely other more appropriate methods to change governments' incentives. The domestic institutional context of a country, namely, the procedural rules governing the budget process, the numerical fiscal rules and independent institutions, if combined with efficient public reporting and performance management policies, can help significantly in fiscal performance.

Based on this experience and until a new system of fiscal governance is established in the EU, we propose a number of basic measures which will at least provide sound conceptual and accounting infrastructure as well as adequate information to the public concerning the disposal of taxpayer money. These measures will have the advantage of (a) helping to improve fiscal performance in general and the "microfoundations" of fiscal policy in particular; (b) being easy to implement; (c) being easily accepted by member countries, (d) fostering social consensus; and (e) consequently, and most importantly, creating the right incentives for governments.

References

Alesina, A., and Tabellini, G. (1990). A positive theory of fiscal deficits and government debt. *The Review of Economic Studies, 57*(3), 403–14.

Audit Commission (UK). (2010). Comprehensive performance indicators. Available at www.audit-commission.gov.uk/cpa/, accessed 8 December 2010.

Baldwin, R., Gros, D., and Laeven, L. (2010). *Completing the Eurozone rescue: What more needs to be done?* London: CEPR.

Blanchard, O. J., and Giavazzi, F. (2004). *Improving the SGP through a proper accounting of public investment*. Discussion Paper 4220. London: CEPR.

Buiter, W., Corsetti, G., and Roubini, N. (1993). Excessive deficits: Sense and nonsense in the Treaty of Maastricht. *Economic Policy, 16*, 57–101.

Buiter, W., and Grafe, C. (2002). *Patching up the Pact: Some suggestions for enhancing fiscal sustainability and macroeconomic stability in the enlarged European Union*. Discussion Paper 3496. London: CEPR.

Caddy, J., and Vergez, C. (2001). *Citizens as partners: Information, consultation and public participation in policy-making*. Paris: OECD.

Coeuré, B., and Pisani-Ferry, J. (2005). *Fiscal policy in EMU: Towards a Sustainability and Growth Pact?* Bruegel Working Paper 2005/01.

Easterly, W. (1999). Fiscal illusion: Taking the bite out of budget cuts. *Economic Policy,* April 1999, 56–76.

Eurostat (2004). *Pensions: Eurostat communication to the Advisory Expert Group.* 2 December.

Ferreiro, J., Fontana, G., and Serrano, F. (2008). *Fiscal policy in the European Union.* Basingstoke: Palgrave Macmillan.

Gleich, H. (2003). *Budget institutions and fiscal performance in Central and Eastern European countries*. ECB Working Paper 215.

Gramberger, M. R. (2001). *Citizens as partners: OECD handbook on information, consultation and public participation in policy-making*. Paris: OECD.

Herzlinger, R. E. (1996). Can public trust in nonprofits and government be restored? *Harvard Business Review*, *74*(2), 97–107.

Koen, V., and van den Noord, P. (2005). *Fiscal gimmickry in Europe: One-off measures and creative accounting*. Working Paper 417. Paris: OECD.

Kopits, G., and Symanski, S. (1998). *Fiscal policy rules*. IMF Occasional Paper 162.

Krogstrup, S., and Wyplosz, C. (2009). Dealing with the deficit bias: Principles and policies. In J. Ayuso-i-Casals, S. Deroose, E. Flores and L. Moulin (Eds.), *Policy instruments for sound fiscal policies: Fiscal rules and institutions*, 23–50. Basingstoke: Palgrave Macmillan.

Lee, M. (2004). *E-reporting: Strengthening democratic accountability*. Washington, DC: IBM Center for the Business of Government.

Lee, M. (2005). Public reporting. In J. Rabin (Ed.), *Encyclopedia of public administration and public policy*, 239–43. Boca Raton: Taylor & Francis.

Milesi-Ferreti, G. M., and Moriyama, K. (2004). *Fiscal adjustment in EU countries: A balance sheet approach*. IMF Working Paper WP/04/143.

Oksanen, H. (2004). *Public pensions in the national accounts and public finance targets*. CESifo Working Paper 1214.

Poterba, J., and Rueben, K. (2001). Fiscal news, state budget rules, and tax-exempt bond yields. *Journal of Urban Economics*, *50*(3), 537–62.

Schiavo-Campo, S., and Tommasi, D. (1999). *Managing government expenditure*. Manila: Asian Development Bank.

Schick, A. (2004). Fiscal institutions versus political will. In G. Kopits (Ed.), *Rules-based fiscal policy in emerging markets: Background, analysis and prospects*, 81–94. Basingstoke: Palgrave Macmillan.

Treasury Board of Canada Secretariat. (2010). Management Accountability Framework, Available at www.tbs-sct.gc.ca/maf-crg/index-eng.asp, accessed 12 December 2010.

von Hagen, J. (1992). *Budgeting procedures and fiscal performance in the European Communities*. European Economy Economic Papers 96.

von Hagen, J., and Harden, I. (1995). Budget processes and commitment to fiscal discipline. *European Economic Review*, *39*(3–4), 771–9.

von Hagen, J., and Wolff, G. (2004). *What do deficits tell us about debt? Empirical evidence on creative accounting with fiscal rules in the EU*. Deutsche Bundesbank Discussion Paper 38.

Wyplosz, C. (2002). *The Stability and Growth Pact: Time to rethink*. Briefing Notes to the Committee for Monetary and Economic Affairs, European Parliament.

Ylaoutinen, S. (2004). *Fiscal frameworks in the Central and Eastern European countries*. Finnish Ministry of Finance, Discussion Paper 72.

Economic Recession and Labour Migration

Helene Mandalenakis

The recent economic crisis is considered to be the most severe there has been in the post–Second World War period. The developed world saw its financial security shattered as large financial institutions collapsed and stock markets began to shake. According to the Business Cycle Dating Committee of the National Bureau of Economic Research, the economic recession began in the United States in December 2007 and ended in June 2009 (NBER, 2010, p. 1). Professor R. J. Gordon, a member of the committee, identifies this recession as the longest and deepest since the Great Depression in terms of job losses (Rambell, 2010). During this period, the United States' economy reached its "low point" and recovery only began in June 2009 (NBER, 2010, p. 1). In the euro area, according to the Business Cycle Dating Committee of the Centre for Economic Policy Research, the recession lasted from January 2008 to April 2009 (CEPR, 2010, pp. 1–2). Unemployment rates are susceptible to economic shocks, the labour market and state responses to the economic crisis. Unemployment in the OECD rose to 8.8% in the fourth quarter of 2009, resulting in an additional 18 million unemployed people (OECD, 2010, p. 84). This crisis reminded policymakers that the economic foundations of the most advanced states were not as secure as was believed, and that they should be strengthened. Once the economies were shaken, they nearly crumbled. Although the crisis affected mainly the more developed states – members of the OECD – its impact was, and still is, felt globally. Economic recovery has been very slow, as governments fear stimulating the economy through public spending. Instead, many European states have adopted austerity measures to decrease their public debts, fearing further economic slowdown.

As expected, the effects are not restricted to the economic sphere but have also affected the social and political sphere. Citizens of the industrialised world watched the economic gains of a lifetime vanish and saw their social rights put aside in favour of unpopular austerity measures that promise economic recovery and a return to prosperity. Social unrest is heightened by demonstrations, strikes in all sectors and even violence against politicians. In Europe, a severely hit region, reactions range from strong social protests (in France) to numbness (in Greece).

Social unrest leads to a search for scapegoats, and in most countries these are the immigrants (Cochrane and Nevitte, 2007). During periods of reduced economic growth immigrant workers are the first to face hostility from the locals because they are blamed for job scarcity. In reality, immigrants accept jobs with lower pay, lower benefits and less job security, jobs that locals do not want.[1] Economic hardship and poverty create fertile ground for rising rates of crime, racism and xenophobia (Awad, 2009, p. 48). Increasing xenophobia leads many to believe that this economic crisis will end immigration and push immigrants back to their countries of origin. If we look only at the economic aspects of migration, this is a reasonable expectation, but how accurate is it?

Economic analysts and financial specialists have been analysing the causes and the depth of the recent recession as well as potential solutions. The goal of this article is to explore the way the economy is linked to international migration. There is no doubt that economic inequality among states induces migration towards the most prosperous economy. It is also expected that states going through economic hardship will not be the preferred migration destinations. This article questions the extent to which economic factors determine and even regulate migration. The argument is that the state is a powerful actor in this process and should not be dismissed. Furthermore, the personal judgement and determination of migrants should be taken into account in order to understand migration flows.

Economic considerations are necessary but insufficient to determine the volume or direction of migration flows. The economy is only one of the

[1] In the United States, low-skilled illegal immigrants are employed in agriculture, as natives are not willing to take these jobs even during harsh economic periods. State policy favours labour importation from other states rather than legalising the Mexicans who have already acquired the skills needed for the job (Westneat, 2010).

determinants of this process. As Hollifield (2004, p. 905) explains, "migration is both a cause and a consequence of economic and political change." The economic and political situation in both sending and receiving countries determines both the volume and direction of international migration. Immigration is both an endogenous and an exogenous process for the receiving country (Sassen, 1999, p. 136). Economic and political conditions in labour-sending countries function as "supply-push" factors and then immigration becomes exogenous to the receiving state. It falls under the receiving state's jurisdiction and immigration policy to reject or accept these immigrants. The accommodation and absorption ability of the labour-receiving state acts as a "demand-push" factor and increases immigration.

State Responses to the Economic Crisis

Policymakers design state immigration policy, always taking into account economic aspects and other socio-political issues. The purpose of immigration policy in liberal democracies is not only to protect or promote the economic interests of a country but to protect and promote the social well-being of its citizens. State security is another aspect of this policy. Hence, immigration policies are shaped by economic, socio-political and even foreign-policy considerations (Cornelius et al., 1994; Sassen, 1999).

Immigration policies vary from country to country according to political ideology, economic development and demographic needs. In particular, "settler" states such as Canada, Australia, New Zealand and the United States, created mainly through immigration, value foreigners' attributes and invite immigration. On the other hand, many European states identifying with a particular ethnicity are more hesitant to accept large numbers of immigrants, fearing that changes in the ethnic composition of their populations will weaken their ethnic or state identities.[2] In both cases, however, there have been exceptions based on the projected needs of the nation state. The phenomenon of migration takes place at different levels,[3] and states try to obtain economic advantages from the movement of labour.

---

[2] In most cases, policies on the socio-political integration of immigrants emanate from the same ideology that determines the openness of the immigration system.

[3] It can be internal, intra-regional or international migration.

Economic development in the settler and the most advanced states in Europe has been achieved through the importation of foreign labour. Immigrants are integral to settler states. European states have a different history. European governments designed "guest worker" recruitment programmes to reinforce the post-war reconstruction of Europe. Each state defined its special needs in terms of labour skills and the length of the work permits. The temporary unskilled or low-skilled workers were expected to return to their country of origin as soon as their work permit expired. This expectation proved to be unrealistic. Temporary workers did not leave as expected but became permanent residents and were eventually naturalised. Furthermore, they invited their families to join them in the host country. This population increase restored the demographic imbalance caused by the wars and strengthened European economies by developing sectors of the economy that had been destroyed.

European states realised that their "guest worker" programmes were bringing not just temporary workers but new citizens, so they chose to halt immigration. During the '70s, European economies were stretched due to the oil crisis and the economic recession that followed (1973–1975). However, the halt to immigration in 1974 was not solely the outcome of these economic events but of the realisation that European identity was changing. Although labour migration officially ended, these states were unable to stop immigration resulting from family reunification.

State reactions to the latest economic recession are policies that give priority to the employment of native-born workers over foreign-born temporary workers. The governments of Singapore and Malaysia are urging companies to first lay off foreign workers. South Korea prefers to subsidise companies that hire nationals (Abella and Ducanes, 2009, p. 8). It is expected that citizenship weighs more and has a positive effect on employment. Governments have to calculate the political cost of the increased unemployment of their electorate.

In reaction to the economic downturn, most countries also stopped or reduced new admissions of foreign workers. Some Asian destination states have issued a freeze on new work visas while Asian sending states such as Sri Lanka and Bangladesh are providing immediate support to their nationals working abroad. They prefer to help them find a new job in the host state rather than to repatriate them. It is interesting to note that although these restrictive measures were enforced as a response to the economic crisis, the governments announced their intentions before the crisis (Abella and Ducanes, 2009, pp. 9–10).

It seems that the recession gave an opportunity to the United States to restrict admission for both low-skilled and high-skilled workers. President Obama also restricted companies receiving government bailout money from hiring H-1B high-skilled workers (Fix et al., 2009, p. 59). Australia also adopted restrictive policies. Canada, in 2009, going against the tide, reduced neither the number of temporary workers permits nor permanent immigration. Since 2007, migrant inflow of low-skilled labour has increased because it is regulated through the Temporary Foreign Workers Programmes (Fix et al., 2009, p. 59; Elgersma, 2007). New Zealand is another country that despite the crisis increased the intake of workers through the Skilled Worker Programme and extended its Skilled Migrant Policy. New Zealand's GDP growth declined in 2008 but state policy favoured immigration increases to offset this decline. New Zealand's policy is to encourage high-skilled immigration during periods of high unemployment (Ongley and Pearson, 1995, pp. 767–70, 787).

Since the late '70s, Europe has been less open to foreign labour migration. It should be taken into account that due to the freedom of European Union (EU) citizens to move and work everywhere in the EU, member states can only regulate the intake of third-country nationals. Despite the adoption of the Blue Card in 2009 (EC, 2009), for many states high-skilled labour immigration remains under national jurisdiction. Among the member states, Sweden is the only state that has liberalised its policy (Cerna, 2010, p. 9).

A reduction in temporary work permits does not necessarily lead to less immigration. In order to boost the economic recovery of certain sectors, states may issue more permits that lead to permanent residency. Hence, states prohibit certain immigration categories but continue to encourage others. Closing the borders to labour migrants may not be a good option in the long run, as it will be harder to bring in workers when the state needs them (Cerna, 2010, p. 23). Also, there is the danger of stimulating illegal immigration through the reduction of legal labour immigration.[4]

---

[4] Liberal developed states are not always efficient in controlling illegal migration due to individuals' determination to defy laws and barriers to reach their destination. Spain and Portugal are good examples of states fighting illegal migration strategically. They defined their strategic interests and through intra-

Another remedy to labour surplus during rising unemployment is to encourage the emigration or return migration of foreign workers. Receiving countries (Spain, Czech Republic, Japan, Greece) and sending countries (Philippines, Argentina, Colombia, Tunisia, China) have now implemented return policies. During a crisis, destination states would encourage the return of migrants in order to alleviate some of the burden of excess labour and decrease unemployment. As a result, some return programmes provide migrants with return cash bonuses and free tickets (Cerna, 2010, pp. 11, 23; Onishenko, 2010).

Feld (2000) challenges the view that industrialised states with demographic problems need to import labour from abroad. He demonstrates that European states need not rely so much on foreign labour as there will not be a shortage of labour until 2020.[5] Taking into account the continuous admission of temporary immigrants as well as the large immigrant communities in most developed countries, Feld's outcomes could still be valid. On the other hand employers, in the United Kingdom and in the United States, for example, are pressuring governments to increase the quota for skilled workers so that companies can find the qualifications they need to become more competitive in international markets (Ahmed, 2010; Manpower, 2010).

A brief examination of state responses to the recent economic crisis reveals that states regulate their intake of labour differently. Political and economic considerations coexist. Consequently, economic pressures do not automatically restrict labour migration. Juan Somavia (ILO Director-General) said that "governments should not have to choose between the demands of financial markets and the needs of their citizens. Financial and social stability must come together. Otherwise, not only the global economy but also social cohesion will be at risk" (ILO, 2010a). It is evident that political and economic interests regarding labour migration do not coincide but diverge, and the cost is significant.

state cooperation (between the sending and receiving states) and efficient policing have effectively protected their borders.

[5] He examined the real need for foreign workers in the twelve original European Union (EC-12) members.

The Economy and Immigration Categories

Economic pressures disproportionately influence different immigration categories. Labour migration is the category most sensitive to economic shocks and downturns. The degree of its vulnerability to economic pressures has already been discussed.

Migration for the purposes of family unification is not directly linked to the economy, as it is justified by the respect for migrant rights. In the late '60s and early '70s immigration policies gave more emphasis to the right to family unification. Among newcomers, preference was given to relatives of established immigrants who could guarantee to provide for their family.

Migration occurring for humanitarian reasons is also not directly linked to the economy. Refugees and asylum seekers relocate due to fear of persecution in their home country because of a civil war or political turmoil. This is forced migration, as most of the refugees and asylum seekers have already been victims of social and political conflict. Signatory states of the Geneva Convention of 28 July 1951 and the Protocol of 31 January 1967 Relating to the Status of Refugees are obliged to accept people who can prove they should be protected by the Convention.[6] The Charter of Fundamental Rights of the European Union, passed in 2000, reaffirms the previous documents.

Migration of EU nationals is not part of the scope of this work because member states cannot restrict intra-Union migration, as freedom of movement and work within the EU is guaranteed by the treaty establishing the European Union.[7]

---

[6] Mounting complaints against certain states (i.e., the United States) refer to the illegal detention of refugees after confirmation of their identity or before their removal due to illegal arrival. Other complaints expose the inhumane living conditions of the detention centres (i.e., in Greece).

[7] States introduce yearly quotas for admission under such programmes as family reunification and based on the applications for humanitarian reasons such as asylum. They have also limited the scope of these categories and have introduced fast-track examination processes of applications. This practice has been implemented irrespective of the economic situation.

Labour Migrants in the Face of the Economic Crisis

Labourers consider emigrating from their home country only when there is economic inequality between their country and the destination country. Labour exchange does not take place between countries with similar economic development. Emigration is viewed as a temporary move to raise money and support the family back home. The goal is to return after having accumulated sufficient capital to help the returnees re-establish in their country of origin.

During a global economic downturn, potential migrants may postpone their emigration until the economy of the destination state has revived. The assumption is that there is an economic motivation to emigrate even during an economic crisis. If, for various reasons, emigration cannot be postponed, it is likely that the individual will reconsider his or her choice of destination to a state with a more stable economy. A supportive social network in the receiving country can greatly facilitate immigrant adjustment to the new society, provide information and guidance to the job seeker and even recommend the immigrant to local employers.

Recent and well-established immigrants in a foreign country may consider leaving if they are unable to integrate into the host society and market. If, however, the cost of relocation is high, leaving may not be an option.

The less integrated migrant workers are into the market, the more sensitive they are to economic downturns.[8] The worker's position in the economy and the state's capacity to deal with the economic downturn determine the severity of the economic crisis for the immigrant. Immigrants are overrepresented in less-skilled occupations. They tend to work with temporary contracts and therefore become subject to discrimination. As a result, they face higher rates of unemployment, earlier layoffs and fewer possibilities of being re-employed. The majority of low-skilled migrants work in industries such as manufacturing and construction, which went through major restructuring during this recession (OECD, 2009, p. 5). It is not easy for low-skilled workers to take advantage of job offers in non-declining industries such as the service sector, simply because they do not qualify for

[8] Examples of these temporary workers groups are the Mexicans in the United States, the North Africans in Spain and the Pakistanis in the United Kingdom (OECD, 2010, p. 95).

them. As a result, they are excluded from the labour market for greater lengths of time.

High-skilled temporary workers have an advantage over the low-skilled. As members of the older generation retire, there are not enough skilled workers to replace them.[9] Employers try to fill this shortage from abroad. The financial sector was also hard hit during this crisis, thus affecting highly skilled workers. In Europe during the 2007–9 period, 363,000 jobs were lost, of which 114,000 were those of immigrants. The numbers are similar for the United States (OECD, 2010, p. 97). It should also be noted that the crisis has greatly affected young foreign-born workers. Since 2009 the United States, Canada and the EU-15 had a young foreign-born unemployment rate of 15.3%, 20.2% and 24.1%, respectively (OECD, 2010, p. 93).

Residence in the host country is determined by the revitalising ability of the economy and the belief that the crisis is only a temporary phenomenon. In fact, many workers choose to "overstay" after the expiration of their work permit, as work opportunities may still be better than in the country of origin (Abella and Ducanes, 2009, pp. 10–11). They can also turn into irregular workers despite increased government crackdown on illegal immigration.

Return migration becomes a choice for well-integrated labour workers under the following conditions: (a) there is a personal need to return; (b) they have attained their economic goals; (c) they can import their skills and take advantage of economic opportunities in their home country; and (d) they can count on a good social network to help them start over.

The likelihood of return declines as years of residence in a foreign country increase. Hence, return migration is most likely to take place within the first three years of migration, while after the fifth year the likelihood of return is very low (OECD, 2008, p. 203). Within the first five years 20–50% of immigrants return to their country or go to another country (OECD, 2008, p. 163).

Stimulus packages from the sending country may persuade migrants to return and use their acquired skills, if any.[10] Return packages from the

---

[9] Usually the acquisition of skills must be accredited. Some countries (e.g., Canada) do not recognise credentials acquired in another country.

[10] Skill acquisition is not always the outcome of working in a more advanced economy. It is determined by the position held by the immigrant. For more, see Hammar et al. (1997, pp. 136–37).

receiving country may also encourage workers to return as they decrease the cost of moving. This measure does not restrict migrants from returning to the destination country when economic conditions permit. Migrants' options are driven by economic concerns and influenced by state policies at both ends. Migrants tend to adjust their objectives to take advantage of favourable immigration policies.

Conclusion

In most states, the economic recession still continues and its effects have not yet been fully evaluated. Economic recovery seems to be taking longer than initially expected so the economic effects on international migration are still unfolding. It is predicted that a return to pre-recession employment will not take place earlier than 2015 (ILO, 2010b). Hence, it should be expected that the volume of temporary migrants will further decline. It is incorrect to consider that migration flows will end, as there are types of migration that are not directly linked to the economy. As economic inequality and political instability continue to exist, migration will remain an option. Illegal immigration will also continue to rise.

Economic downturns can better explain temporary migrant mobility but not migration as a whole. However, the market does not automatically regulate migration; it does so only partly. States respond to economic or development needs and politically regulate the direction and volume of migration inflows accordingly. A political decision can be justified by economic figures. States can reduce immigration just as they can increase it, by creating economic opportunities that will help a weak economy develop. It is a matter of political will and not just economic circumstances. Consideration of many parameters – political, social and cultural, not just economic – contributes to political decision-making. States are also bound by past policies, justified through political and not just economic reasoning. Growing respect for and protection of immigrant rights restrict the state's political power regarding immigration. On the other hand, certain governments may exploit rising anti-immigrant feelings and proceed with otherwise unpopular measures. In both cases, political decisions regarding immigration can jeopardise national unity as a result of varied civic reactions.

References

Abella, M., and Ducanes, G. (2009). *The effect of the global economic crisis on Asian migrant workers and governments' responses*. Bangkok: International Labour Organization (ILO).

Ahmed, K. (2010). Immigration cap damaging the UK, warns services group. *The Telegraph*, 3 October. Available at http://www.telegraph.co.uk/finance/businesslatestnews/8038792/Immigration-cap-dama, accessed 4 October 2010.

Awad, I. (2009). *The global economic crisis and migrant workers: Impact and response.* Geneva: International Labour Organization (ILO).

CEPR (Centre for Economic Policy Research). (2010). Euro area Business Cycle Dating Committee: Determination of the 2009 Q2 trough in economic activity. 4 October, 1–11. Available at http://www.cepr.org/data/dating/Dating-Committee-Findings-04-Oct-2010.pdf, accessed 23 October 2010.

Cerna, L. (2010). Policies and practices of highly skilled migration in times of the economic crisis. International Migration Papers, No. 99. Geneva: International Labour Organization (ILO).

Cochrane, C., and Nevitte, N. (2007). Support for far-right anti-immigration political parties in advanced industrial states, 1980–2005. *European Consortium for Political Research*, 6–8 September. Available at http://www.essex.ac.uk/ecpr/events/generalconference/pisa/papers/PP480.pdf, accessed 6 October 2010.

Cornelius, W. A., Martin, P. L., and Hollifield, J. F. (Eds.). (1994). *Controlling immigration: A global perspective.* Stanford: Stanford University Press.

Elgersma, S. (2007). Temporary foreign workers. Library of Parliament, PRB 07-11E.

EC (European Council). (2009). Council directive 2009/50/EC of 25 May 2009 on the conditions of entry and residence of third-country nationals for the purposes of highly qualified employment. *Official Journal of the European Union*, 18 June.

Feld, S. (2000). Active population growth and immigration hypothesis in Western Europe. *European Journal of Population, 16*(1), 3–40.

Fix, M., et al. (2009). *Migration and the global recession.* Washington, DC: Migration Policy Institute and BBC World Service.

Hammar, T., Brochman, G., Tamas, K., and Faist, T. (Eds.). (1997). *International migration, immobility and development.* Oxford, New York: Berg.

Hollifield, J. F. (2004). The emerging migration state. *International Migration Review, 38*(3), 885–912.

ILO (International Labour Organization). (2010a). Press release, 30 October. Available at http://www.ilo.org/global/About_the_ILO/Media_and_public_information/Press_releases/lang--en/WCMS_145182/index.htm, accessed 20 October 2010.

ILO (International Labour Organization). (2010b). *World of work report 2010: From one crisis to the next?* Geneva: International Institute for Labour Studies.

Manpower (2010). Strategic migration – a short-term solution to the skilled trades shortage. *World of Work Insight*, August. Available at http://files.shareholder.com/downloads/MAN/1042534141x0x397650/5391d32c-1d9e-44eb-a4f2-094d4585eef3/mp_wow_skilled_trades_migration_final_US%20letter.pdf, accessed 20 October 2010.

NBER (National Bureau of Economic Research). (2010). Cycles, 20 September, 1–2. Available at http://www.nber.org/cycles/sept2010.pdf, accessed 23 October 2010.

OECD (Organisation for Economic Co-operation and Development). (2008). Return migration: A new perspective. *International migration outlook: SOPEMI*. Paris: OECD.

OECD (Organisation for Economic Co-operation and Development). (2009). International migration: Charting a course through the crisis. Policy brief. Available at http://www.oecd.org/dataoecd/10/24/43060425.pdf, accessed 3 October 2010.

OECD (Organisation for Economic Co-operation and Development). (2010). *International migration outlook: SOPEMI*. Paris: OECD.

Ongley, P., and Pearson, D. (1995). Post-1945 international migration: New Zealand, Australia and Canada compared. *International Migration Review, 29*(3), 765–93.

Onishenko, K. (2010). Many want to return to their country (in Greek). *Kathimerini*, 14 October, 3.

Rambell, C. (2010). The recession has (officially) ended. *New York Times*, 20 September. Available at http://economix.blogs.nytimes.com/2010/09/20/the-recession-has-officially-ended, accessed 5 October 2010.

Sassen, S. (1999). *Guests and aliens*. New York: The New Press.

Westneat, D. (2010). The fruits of our labour absurdity. *The Seattle Times,* 25 May. Available at http://seattletimes.nwsource.com/html/dannywestneat/2011952772_danny26.html, accessed 23 October 2010.

The Economic Recession in Greece

Ross Fakiolas

Introduction

Intense economic fluctuations in output and overall prosperity have occurred for a long time, with different causes by period and geographic region. The main causes include climate change, pandemics, depletion of basic natural resources, application of new technologies, overproduction of some basic products (transport equipment, houses) and the unwarranted growth of the money and financial sector (Schumpeter, 1943).

The current crisis has been blamed on the excessive expansion of the US derivative market and other products of the financial sector, the high confidence in the ability of markets to self-adjust and the imbalances in the global money and capital markets. The implicit views of the decision-makers were that markets and economies are inherently stable and only temporarily get off track (Lawson, 2009). The recession started in August 2007 in US financial institutions, just as the 1929 economic crisis did. It soon extended to many countries, reducing GDP and increasing unemployment. The global growth rate fell to about 1% in 2009, from about 4% before; and in most countries GDP declined by 3–6%, in some up to 10%. Its duration and long-run effects are still unknown. Annual fiscal deficits of 10% or more and public debts above 80% of GDP, caused by the recession, have turned it into a public debt crisis.

The timely and to a considerable extent well-coordinated neo-Keynesian monetary and fiscal counter-cyclical policy implemented by most governments contributed to reducing the risk of systemic collapse. The policy also maintained an adequate effective demand in order to avert the deepening of the crisis and warded off destitution and social turmoil in any individual country, allowing a weak turnaround in most economies two years after its onset (European Commission, 2010). Vulnerabilities remain, however, leaving no room for complacency and high expectations. The G-20[1] advocates continuing the counter-cyclical measures until the recovery strengthens and gradually shifting the main emphasis to growth.

The Weak Ongoing Recovery

Despite the current growth in global GDP of about 3% on an annual basis and the positive albeit low growth rates in the large economies of the US and Germany (around 1.5%), the recession is not over. In the severely hit developed countries the ongoing recovery is weak, fragile and still depends heavily on state financial support. The fears of a double dip recession, considered unfounded early in 2010, have increased. The Eurozone average for fiscal deficits and public debt stands at about 6% and 82% of GDP, respectively. In some EU-27 countries the debt exceeds 100% of GDP, requiring up to 20% of GDP for annual servicing (nearly one-quarter of it for interest) and increasing further the amortisation-to-GDP ratios. In the increasingly tight global financial markets many countries face rising interest rates (spreads) and credit default swaps (CDS). Even if the recovery continues, it would take many years of strict austerity policies for those countries to reduce their debt to under 60% of GDP, as required by the EU Stability and Growth Pact (SGP).

In addition, unemployment in many countries has soared to over 10% of the labour force, heavily taxing social cohesion, while around 20% of total employment is temporary or part-time work. Long periods and systematic efforts are also required to bring unemployment near 3%, the frictional unemployment rate.

The global averages cited above conceal notably impressive growth rates. China ranks second in global output and first in world trade, ship-

---

[1] The G-20, a new, enlarged group of countries from all continents, has become a unified legal guiding framework for the surveillance of banks, insurance companies and public services. Its purpose is to achieve coordination and cooperation among regulatory authorities for the financial institutions in each country. Under its influence international synergies are enhanced for encouraging social responsibility in the business world and increasing the output of new products with lower production, environmental and running costs.

building, the consumption of energy and the annual registration of new cars (about 12 million), and is recording annual rates above 9%; and India is performing equally well. Therefore, a gradual shift of global economic and political power is taking place, away from the North Atlantic area to the BRICs (Brazil, Russia, India, China). With about half of the earth's population, nearly one-quarter of global output and an appreciable share of advanced technology, the BRICs are changing the world into a multipolar economic and political system. They are also pressing for an international monetary system more efficient to meet present day needs than the Bretton Woods Agreement of 1944. At that time the US generated over half of global GDP, against less than a quarter now. In addition, the BRICs appear increasingly reluctant to lend more to the developed economies (China holds about $800 billion of US government bonds), preferring instead the IMF. They have increased their financial contributions to the IMF and participate actively in its management.

The Recession in the Greek Economy

GDP Decline

In 2007 GDP grew by 4.5% and in 2008 by 2% (versus zero in the Eurozone), declining by about 2.0% in 2009. For 2010 and 2011 the forecasts are for declines of about 4% and 2%, respectively.

Certain country-specific cyclical and structural characteristics acted as buffers to global contagion, allowing the Greek economy to maintain a positive growth differential vis-à-vis the euro area until early 2009. These included support for real wage gains, reflecting a state policy to trade off growth and social peace at the expense of an increase in public debt; lower susceptibility to fluctuations in international trade compared with more industrialised countries, attributed also to the high degree of economic self-sufficiency (currently exports plus imports as a percentage of GDP stand at 48%, against a European average of 70%); a significant deceleration in imports, mitigating the negative impact on domestic GDP of reduced exports; channelling a considerable part of Greece's exports outside the euro area, mostly in emerging countries not affected seriously by the recession; the strong automatic financial stabilisers (social benefits, legal restrictions on dismissals); and the timely application of some counter-cyclical polices, like the state guarantee in 2008 of bank deposits and the lending of €23 billion to the banks (OECD, 2009).

Fiscal Deficit, Public Debt and Inflation

In 2009 the fiscal deficit climbed to 12.7% of GDP and the public debt increased to 113%, from 99.2% in 2008, having acquired internal momentum in its upward trend. The debt is forecast to reach 140% by 2013 because of the high servicing cost, the continuing high fiscal deficit and the inclusion of the payable arrears of local governments, hospitals and the social security funds, as provided for by the EU/IMF agreement. The state owes its suppliers about €10 billion, proportionally higher than in most other countries (€11.5 billion in Germany). Payment takes 157 days compared with 30 in most other countries, and suppliers charge higher prices to compensate for the delays, further damaging state credibility.

The swelling public debt dramatically increased the spreads and CDS, revealing market perceptions of the high risk surrounding future economic developments. The annual servicing cost increased to over €55 billion in 2010 (5.3% of GDP, compared with 5.0% in 2009 and 4.6% in 2008) after a continuous downward trend in the 14 years up to 2007 (Eurobank EFG, 2009; National Bank of Greece, 2010). That cost narrows down the possibilities for fiscal counter-cyclical measures and social care to those affected by the scheduled reforms and high inflation. In October 2010 inflation climbed to 5.3% (against under 2% in the Eurozone), fuelled by the adverse base effects of energy prices, the malfunctioning of the free market and the repeated increases in VAT, other taxes and excise duties on alcohol, tobacco, luxury items and fuel. High inflation in conditions of a deep recession is rare in economic history.

Obstacles to Reducing Fiscal Deficits

Greece's entry into the EMU in 2001 covered many systemic weaknesses in the economy, averted currency devaluations and reduced by more than half the percentage of the GDP required for debt servicing. It could not protect Greece, however, from running high fiscal deficits, living beyond its means and lagging behind many other EU member states in implementing necessary reforms.

Nor could the EMU raise economic competitiveness, which was low even before the crisis, for many reasons: low product quality and high labour costs compared to productivity and the non-price factors of limited research and development, slow progress in innovation, limited outward-looking policies and weaknesses in the tax and credit systems (Bank of Greece, 2009). Low competitiveness is reflected in the limited exports of high technology goods and basic raw materials (10% of the total), causing a foreign trade deficit of about 3% of GDP.

With the euro as its national currency, Greece could not implement the classical means of reducing its fiscal deficit – currency devaluation – as many other non–euro area countries have done. To increase competitiveness and gradually reduce the high public debt, Greece has resorted instead to severe austerity measures, hoping to meet its obligations to the European Commission albeit at the expense of intense social reaction.

Borrowing from the EU and IMF

The government hesitated for seven months to implement counter-cyclical policies, until in early May 2010 the cost of CDS surpassed 1,000 ppts (percentage points above the 3% interest rate for sovereign German debt). Under this pressure, the Greek government concluded an agreement with the EU, ECB and IMF (the "troika") for a three-year loan of €110 billion at 5% interest. The loan will be dispersed in instalments upon successful completion of each quarterly review. Additional measures will be requested in case of deviations from the programme's targets. Repayment of the loan will start after the last instalment is dispersed. Detailed terms were specified in a Memorandum voted on by the Greek Parliament, while close monitoring by the EU/IMF limits the scope for government non-compliance.

The 2010–12, renewed SGP targets a reduction of 9.5 percentage points in the state budget deficit, to about 3% of GDP by 2012. For this most ambitious fiscal adjustment in conditions of growing social unrest, an assortment of permanent and one-time non-recurring measures have been taken. They consist of increasing revenue and cutting expenditure, intensified efforts against the evasion of tax and social security contributions, a recruitment and wage freeze in the public sector, a cut in all operational expenses of ministries, a hike in various taxes and many others measures.

Six months later (at the end of November 2010) it has become clear that the reforms have not been front-loaded. The most painful measures for diminishing the size of the broader public sector and changing labour relations in important public utilities have not been implemented. In addition hardly any policy has been implemented for promoting economic growth through increasing continuously declining public investment and attracting more foreign direct investment. The Bank of Greece forecasts further deterioration of the economic situation (Bank of Greece, 2010).

Recently the CDS soared again to over 800 ppts, indicating the misgivings of the financial markets and of many analysts at home and abroad about Greece's ability to meet the terms of the Memorandum. Greek officials and EU/IMF executives appear to discount those reactions; they have stressed repeatedly that there is no question of Greece defaulting on its debt, although an extension of the original deadline for repayment could be negotiated with the creditors.

Effects on Greece of the Recession in the EU

The EU is a large, stable, international entity with strong and growing economic and political influence on world affairs, aspiring to be more than an association of countries with an extended market. It is interested therefore in averting the recent deepening of the recession in Ireland and the Mediterranean Eurozone countries. The ECB contributes by discounting state-guaranteed bank bonds at 1% interest, irrespective of their grading by the international rating institutions. It has reaffirmed that it will continue this policy beyond 2010. Recently, however, it discounted the bonds at their reduced market value, up to 30% below nominal value (so-called haircutting). By the end of October 2010 Greece had received about €95 billion in bond guarantees.

In addition, the EU took a first step towards a "fiscal Europe" in May 2010 by setting up a European Financial Stability Fund of €750 billion to lend to member countries in financial difficulties at low interest rates but on IMF terms. Of the 16 Eurozone countries (Estonia, the 17th member, joins on 1 January 2011), 13 are not in compliance with SGP rules and 4 (Ireland, Portugal, Spain and Greece) are in real financial difficulties.

A second step is perhaps the German-French proposal to impose strict penalties (including the temporary loss of voting rights) on countries failing to comply with SGP rules. The proposal was rejected because, among other factors, a revision of the Lisbon Treaty (1999) would be necessary. The conference however agreed on a limited revision of the Treaty.

The Challenge of Unemployment

The direct drag on disposable income by the fiscal tightening causes a vicious circle: declining public spending reduces the demand for labour, while high unemployment and population ageing necessitate additional state expenses for social insurance and welfare. The risk exists of a civil backlash if conditions in the labour market continue to deteriorate.

Even before the recession the economy was facing high structural unemployment, a rigid labour market and an unsustainable welfare state. Due to an average 4% annual growth rate for about 10 years before 2008, unemployment had declined from 11.5% of the labour force in 2000 to 7.2% in 2008. But it was up again to 12.5% in October 2010 (over 600,000 persons), above the EU average. The OECD emphasises that increasing spending to increase employment (including mobility to high productivity sectors) is more beneficial in comparison with many alternative approaches, even in countries with high fiscal deficits (OECD, 2008)

Due to the rising uncertainty caused by the recession, self-employment (including that of farmers) has increased to above 40%, double the OECD average. Based on necessity, namely, to find employment (usually for other family members as well), productivity in this sector is low. It differs therefore from self-employment based on business opportunities that promise a higher income generated as a rule by innovation and efficiency (Ioannidis and Tsacanikas, 2007). Employees, on the other hand, amount to about 65% of total employment (compared with over 80% in the EU) and most of them aspire to become civil servants.

Difficulties Maintaining Full-Time Employment

Due to productivity gains of about 1.5% annually, which tend to be higher in times of recovery, unemployment increases even with a low growth rate, as happens in conditions of a weak recovery. For this reason the IMF forecasts unemployment in Greece at 15% of the labour force in the coming 4–5 years (IMF, 2010). To verify, however, the real quantitative situation in the labour market would necessitate examining carefully the labour force participation of the population and the degree of flexible employment, which allows more people to participate in the economy (EU Summit Conference, Prague, 7 May 2009). Flexible employment amounts to about 10% in Greece, against an average of 20% in the euro area. The International Organization for Migration (IOM) argues that policies supporting flexible employment change labour relations radically. In a blunter statement, the International Labour Organization (ILO) stresses the necessity of encouraging flexible employment and also scheduling "part-time employment", that is, long periods of unemployment which may last for many months (ILO Conference, Geneva, 3 June 2009). The drawback of this suggestion is that it may increase unregistered employment, especially in less-organised societies with many immigrants, like Greece.[2] Inducing firms to hire unemployed persons via subsidised programmes for employment appears more effective.

Flexibility

Flexibility is a complex multidimensional notion including, first, less than full time and various kinds of employment. Even before the recession, temporary and part-time employment increased in all countries for technological, organisational and social reasons. Second, it includes the pressing issue of flexicurity in conditions of austerity, that is, the need to replenish a considerable part of the income lost through dismissals. A third element is the degree of flexibility of every employee, namely the extent to which employees are willing and able to move to other jobs with high market demand and to avail themselves of existing educational opportunities. It is indispensable that education and training adjust to the requirements of technological progress; but this adjustment requires in addition an efficient state mechanism for flexicurity. In this way flexibility also facilitates the necessary evolution of production from simple agricultural jobs and textiles to the mechanisation of the productive process, the application of electric and electronic devices, nanotechnology and biotechnology.

---

[2] The recent social turmoil over undocumented immigrants originating in war zones in Africa and Asia necessitates more careful study of their role in meeting labour market needs and achieving demographic equilibrium, as well as their effects on local communities.

Policy Issues

The main policy challenges include the reduction of the fiscal deficit in an effort to stabilise first the debt-to-GDP ratio and then to create primary fiscal surpluses (i.e., excluding interest costs). This will lead to a gradual decline of the public debt and the turning around of the still-negative growth rate. It is equally important that economic and social reforms be implemented without delay.

Higher productivity would contribute to a sustainable upturn, based on Greece's significant sources for growth: bountiful state property in shares, bonds, real estate of 60 million stremmata (6 million hectares) and so on, estimated at about the same value as the public debt; increased public revenues through privatisation; capturing a large part of the underground economy; developed entrepreneurship and large annual numbers of tertiary education graduates; increased exports and activity in shipping, following a recovery in the world trade; a larger share in tourism; recruiting more private funds and business initiative through Public–Private Partnerships (PPP) (i.e., increasing PPPs); accelerating the pace of absorbing the remaining €23 billion of the National Strategic Plan for Growth, out of a total of €26.2 billion for the period 2006–13; and utilising the new loan received recently from the European Investment Bank (see also Eurobank EFG, 2010).

Conclusions

The effective counter-cyclical measures implemented by most governments allowed many economies to enter a weak recovery two years after the onset of the recession. However, quantitative easing by the central banks – state financial support – continues, causing large fiscal deficits and growing public debts, while unemployment remains high. As things stand now, a double recession cannot be ruled out.

The declining GDP in Greece is not expected to turn positive before 2012, and a long austerity period is necessary to reduce the debt and increase public investment after many years of decline. In addition, the main political parties have failed to attain the minimum consensus necessary to avoid social unrest. It is encouraging, however, that many reforms have been introduced and others are to follow, while an increasing percentage of the population realises the necessity to lower disposal incomes and continue the scheduled reforms.

References

Bank of Greece. (2009). *Bulletin of conjunctural indicators*. October (124). Available at http://www.bankofgreece.gr/BogEkdoseis/sdos200910.pdf, accessed 15 November 2010.

Bank of Greece. (2010). *Intermediate report*. October. Available at http://www.bankofgreece.gr/BogEkdoseis/Inter_NomPol2010.pdf, accessed 15 November 2010.

Eurobank EFG. (2009). *Greece: Macro monitor. A quarterly review of the Greek economy*. November. Available at http://www.eurobank.gr/Uploads/Reports/EnMacroMonitorNov2009New.pdf, accessed 15 November 2010.

Eurobank EFG. (2010). *Economy and markets*. June. Available at http://www.eurobank.gr/Uploads/Reports/EMGrEconomyStabilityJune302010Final.pdf, accessed 15 November 2010.

European Commission. (2010). *Interim forecast, September 2010*. Available at http://ec.europa.eu/economy_finance/articles/pdf/2010-09-13-interim_forecast_en.pdf, accessed 15 November 2010.

IMF (International Monetary Fund). (2010). Statement. 11 May.

Ioannidis, S., and Tsacanikas, A. (2007) *Entrepreneurship in Greece, 2006–2007* (in Greek). Athens: Foundation for Economic and Industrial Research.

Lawson, T. (2009). The current economic crisis: Its nature and the course of academic economics. *Cambridge Journal of Economics*, *33*, 759–77.

National Bank of Greece. (2010). *Greece: Monthly macroeconomic outlook*. January. Available at http://www.nbg.gr/wps/wcm/connect/2feeef004159e38190c49260e5dd86a2/Monthly_Jan2010_5.pdf?MOD=AJPERES&CACHEID=2feeef004159e38190c49260e5dd86a2, accessed 15 November 2010.

OECD (Organisation for Economic Co-operation and Development). (2008). *Education at a glance 2008: OECD indicators*. Paris: OECD.

OECD (Organisation for Economic Co-operation and Development). (2009). *Economic survey of Greece*. Available at http://www.oecd.org/dataoecd/5/61/43284926.pdf, accessed 15 November 2010.

Schumpeter, J. A. (1943). *Capitalism, socialism and democracy*. London: G. Allen and Unwin.

The Political Economy of the Greek Crisis in the Framework of the European Monetary Union*

Pantelis Sklias

Introduction

In 1991, the Maastricht Treaty was signed in the eponymous Dutch town by all the European member states. This unique project, which was ratified by the last country (Germany) two years later, marked the end of a long road to achieving the European Economic and Monetary Union (EMU). From its earliest beginnings, the European project was a result of intergovernmental bargaining by the most powerful European countries in order to overcome global, regional and national challenges (Moravcsik, 1998; Hosli, 2000). Thus, under these conditions the European Union was strengthened and its common identity was promoted by its most powerful actors.

However, 20 years later the EMU's political, economic and institutional context seems to be a rather complicated framework for certain member states, such as Greece, Portugal, Spain and Ireland – namely, the "weak" economies. The Greek crisis not only revealed but also underscored how the EMU is a complex political and economic environment in which many peripheral European economies remain vulnerable to the political and economic shocks that can happen on a regular basis, mainly because of conflicting national interests and preferences at European level. As De Grauwe (2010b, p. 172) argues, "large areas of economic policies remain in the hand of national governments, creating asymmetric shocks that undermine the sustainability of the monetary union." In this regard, the

* I want to thank George Maris, PhD candidate at the Department of Political Science and International Relations, University of Peloponnese, for his valuable contribution in shaping the main arguments of this paper.

weak European politico-economic institutional context, in which the current Greek crisis has developed, has also been a contributing factor. This context can be better analysed in two different stages: first, by applying Optimal Currency Areas (OCA) criteria to the EMU; and second, by evaluating the main political and economic institutions and their vulnerabilities at European level. As will be shown, many European countries and especially Greece cannot afford the vulnerability and one-sidedness of the weak political and economic European context, in which they lose important elements of their political and economic sovereignty. This context, together with their own political and economic structural deficiencies (Sklias and Galatsidas, 2010), creates an explosive mixture of conditions that in the long run will spill over with a negative impact, regionally and globally.

The Application of OCA Criteria to the EMU

Baldwin and Wyplosz (2006) propose six criteria for examining whether a country is able to participate in a monetary union. These criteria can be based either on economic factors like labour mobility, production diversification and openness, or on political elements like fiscal transfers, homogeneous preferences and the solidarity criterion. According to Baldwin and Wyplosz, European countries do not satisfy either the labour mobility or the fiscal transfer criterion, they partially satisfy the homogeneity of preferences criterion and it is very unclear whether there exists a shared sense of solidarity. Instead, European countries satisfy the trade openness and the production diversification criteria. Moreover, it can also be observed that within the EMU there exist major economic divergences in gross domestic product (GDP), employment, labour productivity, budget deficits and debts. In this regard, as Lapavitsas, Kaltenbrunner, Lindo et al. (2010, p. 5) believe, "the integration of peripheral countries in the Eurozone has thus been precarious, leaving them vulnerable to the crisis of 2007–9 and eventually leading to the sovereign debt crisis." As a result, the partial fulfilment of the OCA criteria shows that participation within the EMU could create many disadvantages and costs for the majority of European countries. Thus, the process of monetary unification can be also explained as a political project based on national preferences and interests.

More specifically, according to the OCA framework many have argued that the United States is in a much better position than the European Union is (Eichengreen, 1990; Bayoumi and Eichengreen, 1992). Now, nearly 20 years later, it is obvious that many critical elements of the OCA theory were disregarded by numerous European political leaders and their states, both weak and strong.

First, as Hix (2005, p. 338) argues, "a monetary union should be able to adapt to asymmetric economic cycles either through labour movement from states in recession to states in high growth, or through reductions in wage and labour costs in states in recession." In this respect, labour mobility in the EU is lower than that it is in the US. This suggests that when an asymmetric shock – that is, in real wages – happens within the euro area, not only is the EU unable to address this shock because of its lack of labour mobility but also that there are significant differences among member states in the means and real ability for action. Thus, Greece is much less equipped to overcome this asymmetric shock than Germany because of the subjective and objective costs to its people of moving away: the differences in culture and language from other European countries are much greater for Greece than for any other European country.[1]

Second, according to the trade dissimilarity index, it is clear that there are significant differences in the effect of trade on monetary integration. Even though Greece is second after Norway in terms of the dissimilarity score (Germany is the reference point), its economy is not as integrated with the European economy (Baldwin and Wyplosz, 2006). Thus, the EMU has not been an advantage in this case, either.

Third, most members of the EMU seem to fulfil the openness criterion while at the same time increasing their average ratio of exports and imports to GDP and large pass-through coefficients. However, again Greece seems to be an exception. According to the *Statistical Annex of European Economy* (European Commission, 2005, p. 186), Greece has the lowest ratio of exports and imports to GDP, at 25.5%. In contrast, Italy's ratio is 27.9%, Portugal's is 36.2%, Cyprus's is 48.3%, Bulgaria's is 65.9% and Turkey's is 36.5%. As a result, it can also be argued that in terms of openness, Greece has been unable to participate in the EMU because it is an extreme example in comparison with other European countries. Again, this

---

1 Thus even now, two years after the Greek crisis, Greeks are not willing to move away and work abroad in order to compensate for the increased costs at home, but seem to be withdrawing from the labour force. For an empirical explanation of this phenomenon see Fatas (2000).

fact has been overlooked not only by Greek but also by European political leaders.

Furthermore, as Bayoumi and Eichengreen (1992, p. 35) had predicted, the fact that "supply shocks are larger in magnitude and less correlated across regions in Europe than in the United States underscores that the European Community may find it more difficult to operate a monetary union than the United States." For them it was obvious that the United States would be able to overcome a shock faster than the EU could. In this respect, the current global financial and economic crisis that started in the US in 2007 proved them correct. However, if this is the case it means that many European countries that do not operate at the centre of the European Union, like Greece, Spain and Portugal, are not only affected by very different economic shocks but also that unlike the core countries they have little ability to surpass these negative shocks. This means that Greece, as was demonstrated after the global crisis in 2007, is theoretically and empirically unable to manage the exogenous economic shocks that were converted to endogenous economic shocks within the EMU.

Institutional Political and Economic Vulnerabilities for Greece in the EMU

The above discussion has made it evident that within the EMU there are significant differences and divergences among countries in the applicability of the first three OCA criteria, differences that directly affect each country's ability to overcome an external shock. The OCA framework provides three more critical factors for evaluating each country's ability to participate in a currency union. However, these last three criteria – homogeneous preferences, fiscal transfers and solidarity – are related not only to European economic performance but also to political and institutional vulnerabilities that can be observed at European level. These political and institutional weaknesses at European level affect the performance of the EMU's member states and especially the performance of the peripheral countries. In this regard, Greece remains helpless within the European context when an economic shock directly or indirectly affects its performance.

First, the homogeneity of preferences criterion means that any time a crisis arises, a monetary union should propose common actions on how to solve it. In this respect, the current crisis and especially the case of Greece has proved that within the EMU there is no common consensus on how to deal with crises. For the last two years, European countries have tried individually to compensate for the negative effects of the global financial crisis. National preferences and interests have prevailed at European level. Moreover, it is clear that at this time Greece needs a very special monetary and fiscal policy that within the EMU is regarded as a utopia. However, this is not the only case, since even in the monetary field, for example, Greece and Portugal have very little in common with Germany and France. The great divergences in GDP, in current account deficits and public debts and in inflation rates between the core European countries and the peripheral countries have caused an explosive political and economic environment within the EMU.

Under the aforementioned conditions the effectiveness and acceptance of European institutions decreases not only because within the EMU there is no homogeneity of preferences but also because these institutions, such as the European Central Bank (ECB), have been developed in order to serve a monetary union with a rather limited sense of solidarity. Thus, the ECB suffers from a lack of transparency, credibility and reputation. According to Hix (2005, p. 330), "without an established reputation, public opinion in states that suffer asymmetric shocks is likely to turn against the ECB quicker than it would against a national central bank … without a binding commitment by and clear incentives for the governments to abide by these contracts, the credibility of these coordination efforts is questionable." Furthermore, it can also be said that anytime the ECB's Executive Board follows the method of "one member, one vote" and the economic cycles among the large and small counties are unequally allocated, then severe conflicts and crises can arise (Bindseil, 2001). Thus, when a sudden economic shock like the Greek crisis influences the Eurozone, the effectiveness of the ECB's policies decreases. It is obvious that the ECB cannot be the best representative financial institution when the EMU includes countries with huge political and economic differences between them, like Greece and Germany. In other words, the ECB is still unable to help deal with each country's challenges, which is another part of the same problem.

On the other hand, the Stability and Growth Pact (SGP) that was adopted as one of the main institutional projects for stability and growth within the Eurozone seems to be a rather poor and unsustainable institution (De Grauwe, 2010b). As McNamara (2005, p. 156) argues, "although the

SGP has the word growth in its title, it is not likely to promote growth, but rather to be excessively restrictive." According to Hix (2005, p. 336), "in practice the pact was not a credible way of coordinating national fiscal policies and European level monetary policies in a monetary union with divergent economic cycles, as governments would always respond to their voter preferences first." In this respect, not only have countries like Greece breached the limits of the SGP many times, but it also seems that at the same time these countries have been trapped under the institutionally weak guidelines of the SGP, a fact that undermines the sustainability of the Eurozone.

Finally, not only is a mechanism for fiscal transfers within the EMU lacking, it is also obvious that the coordination between fiscal and monetary policies has failed. On the one hand, according to Jovanovic (2005, p. 60), in the EU "unlike in the US, there is no fiscal element in the deal. Automatic fiscal transfers as built-in stabilisers do not exist." In this regard, the EU needs an economic institution that can organise not only the monetary but also the budgetary and fiscal policies of the Eurozone (Verdun, 2010). On the other hand, there is strong political resistance from many European countries to the possibility of a coordination mechanism (Hix, 2005). Thus, "a one size fits all monetary policy that cannot accommodate regional variations and lacks adequate mechanism for fiscal transfers will impede rather than promote efficient operation of the internal market" (Gillingham 2003, p. 269). According to De Grauwe (2010a, p. 3), "this imbalance leads to creeping divergencies between member states and there is no mechanism to correct or to alleviate them." The aforementioned deficiencies are important elements of the European and Greek failure. As long as Germany and other core European countries are able to follow their own economic policies, as in the case of relative unit labour costs in the Eurozone (figure 1) and changes in the intra–euro area real effective exchange rate, in which it is obvious that Germany's wage moderation policy was significantly different from that of other member states, then not only will the economic divergences within the euro area continue to exist but also many peripheral countries will be unable to resist, trapped within the euro area.

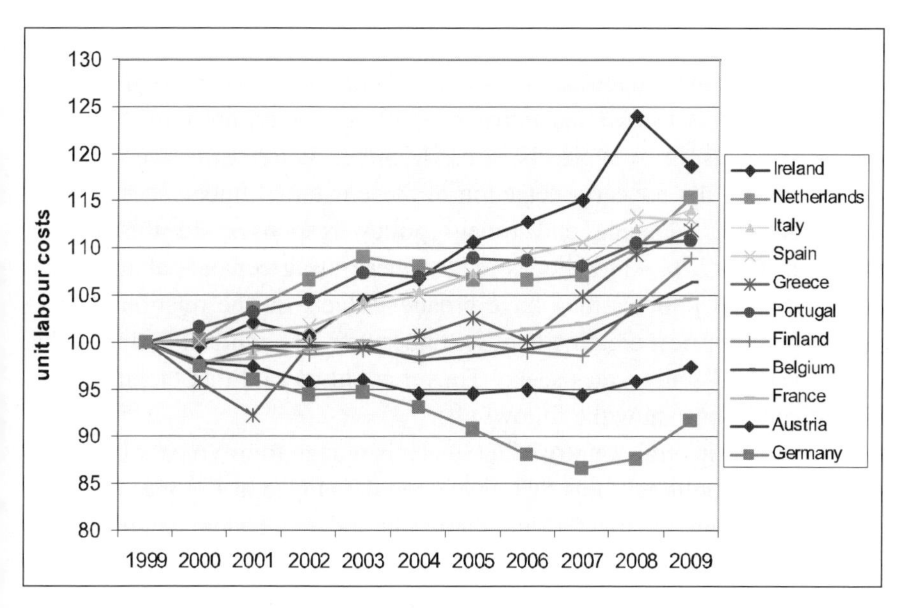

Figure 1. Relative unit labour costs in the Eurozone (Source: De Grauwe, 2010a)

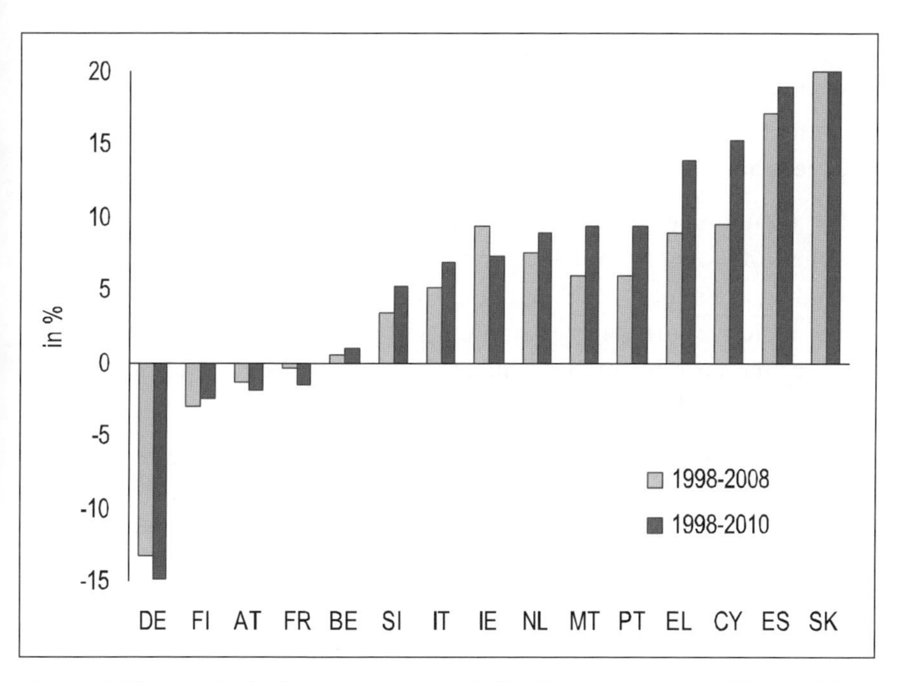

Figure 2. Changes in the intra–euro area real effective exchange rate (Source: European Commission, 2009)

Thus, as the European Commission (2009, p. 35) argues, "Persistent divergence in competitiveness is a matter of common concern as intra–euro area adjustments to external imbalances work slowly, are costly and can have negative spill-over effects across Member States. Effective functioning of EMU calls for early detection of these external imbalances in order to prompt an adequate and timely policy response." In this regard, Feldstein (1997, p. 41) believes that "[t]hese disagreements about monetary and fiscal policies may have broader effects on the relations among European countries, creating conflict rather than the political harmony that many of EMU's advocates seek." Thus it could be said that the aforementioned conditions show the following:

1. A small peripheral country like Greece is unable to use most of the economic and political tools that have been developed at European level in order to overcome many of the regular or sudden economic and political shocks within the Eurozone.
2. Such countries did not initially fulfil the criteria for participation in the EMU. In this respect, it seems that Greece entered the European context without the required qualifications. Eventually, Greece entered into a vicious cycle without the necessary eligibility qualifications.
3. Within the EMU the political and economic environment is violent and conflicted rather than harmonious; weak economies such as that of Greece are exposed to both internal and external shocks.

But the big question remains: Why then did Greece not only decide to join the EMU but also become a member so easily? The answer is politics. According to Frieden (1998, p. 26), three political reasons made joining the EMU attractive: (a) the quest for anti-inflationary credibility; (b) broader links to European integration; and (c) the support from powerful business interests. As Feldstein (1997, p. 41) argues:

> Political leaders in Europe seem prepared to ignore these adverse consequences because they see EMU as a way to further the political agenda of a federalist European political union, which will have a common foreign and military policy and a much more centralised determination of what are currently nationally determined economic and social policies. Although such a policy is often advocated as a way to reduce conflict within Europe, it may well have the opposite effect.

In this regard, it seems that many economic reasons for non-participation within the EMU were ignored by European political leaders, and this behaviour can be explained only in the name of national preferences and interests. In different circumstances not only could coordination within the EMU be achieved but also the European project could be fully realised.

Conclusion

The foregoing discussion has analysed some of the most common weaknesses of the European context that directly or indirectly affect economic and political performance. According to the analysis above it is not completely obvious whether the rules and principles that exist and function within the EMU are concrete ones that create a solid basis for viable economic and financial policies for its members. Instead, it seems that the EMU's structural disadvantages, artificial construction, absence of effective coordination and vague vision remain its most prominent deficiencies (Cohen, 2008; Cohen, 2009). As analysis shows, the EMU from its early beginnings was grounded not on each country's actual ability to participate but on political reasons based on fears of exclusion. In this regard, political priorities have definitely overtaken economic and fiscal principles in the formation and function of the EMU. However, it is obvious that the rules of the game are not viable anymore. Greece and the other peripheral countries cannot afford to remain in a European context full of deficiencies, flaws and inequalities. Either as a matter of economic or political necessity, a great transformation has to occur shortly. And of course Greece must play a role in the process.

References

Baldwin, R., and Wyplosz, C. (2006). *The economics of European integration*. Berkshire: McGraw-Hill Education.

Bayoumi, T., and Eichengreen, B. (1992). Shocking aspects of European Monetary Unification. NBER Working Paper 3949.

Bindseil, U. (2001). A coalition-form analysis of the "one country–one vote" rule in the Governing Council of the European Central Bank. *International Economic Journal*, *15*(1), 141–64.

Cohen, B. (2008). The euro in a global context: Challenges and capacities. In K. Dyson (Ed.), *The euro at 10: Europeanisation, power, and convergence*, 37–53. Oxford: Oxford University Press.

Cohen, B. (2009). Dollar dominance, euro aspirations: Recipe for discord? *Journal of Common Market Studies*, *47*(4), 741–66.
De Grauwe, P. (2010a). Crisis in the Eurozone and how to deal with it. Centre for European Policy Studies, CEPS Policy Brief 204. February.
De Grauwe, P. (2010b). The fragility of the Eurozone's institutions. *Open Economic Review*, *21*, 167–74.
Eichengreen, B. (1990). Costs and benefits of European Monetary Unification. Centre for Economic Policy Research, Discussion Paper 453. London.
European Commission (2005). *Statistical annex of European Economy*. Directorate-General for Economic and Financial Affairs. Spring.
European Commission (2009). *Annual report on the euro area – 2009*. Directorate-General for Economic and Financial Affairs. European Economy 6. Luxembourg: Office for Official Publications of the European Communities.
Fatas, A. (2000). Intranational migration: Business cycles and growth. In G. Hess and E. Van Wincoop (Eds.), *Intranational macroeconomics*, 156–88. Cambridge: Cambridge University Press
Feldstein, M. (1997). The political economy of the European Economic and Monetary Union: Political sources of and economic liability. *Journal of Economic Perspectives*, *11*(4), 23–42.
Frieden, J. (1998). The euro: Who wins? Who loses? *Foreign Affairs*, Fall, 25–41.
Gillingham, J. (2003). *European integration 1950–2003: Superstate or new market economy?* Cambridge: Cambridge University Press.
Hix, S. (2005). *The political system of the European Union*. New York: Palgrave Macmillan.
Hosli, O. (2000). The creation of the European Economic and Monetary Union (EMU): Intergovernmental negotiations and two-level games. *Journal of European Public Policy*, *7*(5), 744–66.
Jovanovic, M. (2005). *The economics of European integration: Limits and prospects*. Northhampton, MA: Edward Elgar.
Lapavitsas, C., Kaltenbrunner, A., Lindo, D. et al. (2010). Eurozone crisis: Beggar thyself and thy neighbour. RMF Occasional Report. March.
McNamara, K. R. (2005). Economic and Monetary Union: Innovation and challenges for the euro. In H. Wallace, W. Wallace and M. A. Pollack (Eds.), *Policy-making in the European Union*, 141–60. Oxford: Oxford University Press.
Moravcsik, A. (1998). *The choice for Europe: Social purpose and state power from Messina to Maastricht*. Ithaca, NY: Cornell University Press.
Sklias, P., and Galatsidas, G. (2010). The political economy of the Greek crisis: Roots, causes and perspectives for sustainable development. *Middle Eastern Finance and Economics*, *7*, 166–77.
Verdun, A. (2010). Economic and Monetary Union. In M. Cini and N. Perez-Solorzano Borragan (Eds.), *European Union politics*, 324–39. Oxford: Oxford University Press.

Is Greece a Failing Developed State? Causes and Socio-economic Consequences of the Financial Crisis*

Harris Mylonas

The Context

Is the Greek crisis an isolated case or the first of a series of future failing developed states? The Greek financial crisis is not on the front page of the *Financial Times* anymore, but it is far from over. The financial crisis did not manifest itself in Greece alone. Ireland has also sought an equally large EU-IMF rescue plan. Portugal and Spain have been under the microscope of the media and credit rating institutions. Such other instances in the Eurozone's periphery have repercussions for the currency as a whole as well as for the EU (Straubhaar, 2010). Greece, Ireland, Portugal and Spain are members of the Eurozone area, which means that they share the same currency with economic giants such as Germany and France.

Greece's debt is primarily owned by French, German and, to a lesser extent, British banks. If Greece defaults, this would severely undermine the confidence in other "risky" countries such as Spain, Portugal and Ireland. Northern European banks have a total exposure of two trillion euros to these countries.[1] Thus, a Greek default could bring down many of these banks – in addition to most of the Greek ones – and ultimately could unravel the whole financial system of the European Union.

* I would like to thank Andreas Akaras, Elias Carayannis, Dimitrios Lagias, Manolis Galenianos, Nick Karambelas, Dimitrios Mantoulidis, Angelos Pangratis, Georgios Skoulakis and Marilena Zackheos for their helpful comments. This paper is based on a talk I gave at the Center for European Studies–European Union Center, at the University of Michigan, Ann Arbor, MI, on 16 September 2010.

[1] This figure includes private debt.

This article aims to first address the deep and more proximate causes of Greece's public debt crisis that include the Ottoman legacy, Greece's geographic location, populism and patronage politics, repeated electoral cycles and endemic corruption. It then discusses a wide range of socio-economic consequences – negative as well as positive – that the financial crisis has had (or will have). The negative consequences include an intensification of conspiracy theories and of the brain drain, high unemployment accompanied by a feeling of hopelessness, the complete bankruptcy of the political system and a reverse migration of the most productive immigrants in the country. The article concludes with a discussion of some developments that can be viewed as opportunities for contemporary Greece with more positive consequences: the "taming" of the public sector, structural reforms, the stabilisation of migratory and refugee flows to Greece and decentralisation.

With other EU member states, such as Ireland and Portugal, having found themselves in a similar situation to that of Greece, albeit through a different path (Legge, 2010), it appears that the problem is also a systemic one. Both Ireland and Greece have resorted to EU-IMF rescue plans in the face of enormous deficits, poor credit ratings and sky-high interest rates to borrow money. Like Greece, Ireland is also at the periphery of the EU and has also borrowed billions from European banks with the possibility of a default. In this sense, Greece is not a unique case of a failing developed state. It also need not be one at all if adequate measures are taken in a timely manner.

The Deep and Proximate Causes

Within the past 180 years, modern Greece has found itself in default-like situations for more than 50 of these years. Most defaults emerged following war: after the War of Independence in the 1820s, at the end of the nineteenth century as a result of the Greek–Turkish war, in the mid-1920s as a result of the Asia Minor Catastrophe and at the end of the Second World War. These default precedents, which many people point to, are not relevant this time around. This time Greece is facing such a situation in the absence of war. Maybe the defaults of 1893 and 1932, the only ones not following a war, are more relevant to the current situation.

There are other historical factors that scholars and commentators have advanced. Some point to the Ottoman legacy as an explanatory factor for Greece's problems (*Hürriyet Daily News*, 2010). This argument follows the logic that the Ottoman legacy is to blame for patronage politics, endemic corruption, populism and nepotism in the country. Populism has thrived in Greece and still does so. In the past two centuries of Greek politics, there have been some exceptions but these have not managed to reverse the trend. Endemic populism has led to vested interests, corrupt practices, a large public sector and a culture of reliance on the state (Pappas and Assimakopoulou, forthcoming).

Robert Kaplan recently blamed Greece's geography for its political and economic troubles. He argues that

> [t]he relatively poor quality of Mediterranean soils favored large holdings that were, perforce, under the control of the wealthy. This contributed to an inflexible social order, in which middle classes developed much later than in northern Europe, and which led to economic and political pathologies like statism and autocracy. It's no surprise that for the last half-century Greek politics have been dominated by two families, the Karamanlises and the Papandreous. (Kaplan, 2010)

Beyond the – debatable – impact that Greece's history of past rule, natural endowments and mode of production have had on its socio-economic development, there is a direct way that geography accounts for Greece's troubles. The country maintains the second highest defence spending level in NATO (next to the US) and the highest in the EU on a percentage basis. This is because of Turkey. This spending and the accounting for this spending is arguably a contributing cause to the accumulating debt.

Yet there are more proximate causes of Greece's present financial crisis. In the past 30 years, Greece has experienced strong electoral cycles, which have increased deficits and built up an enormous state debt over time (Alogoskoufis, 1995). Contemporaneously, the European Community's developmental funding in the early 1980s supported projects designed to solidify the political base of the ruling party. Andreas Papandreou pioneered this practice (Lyrintzis, 1987; Mavrogordatos, 1993, 1997). The end result of such patronage politics and the need to service the public debt kept resources from being invested in productive activities that could lead to sustainable growth (Meghir, Vayanos and Vettas, 2010). There were attempts in the 1990s to reverse this trend, but to no avail.

Paradoxically, another proximate cause was Greece's entry into the European Economic and Monetary Union (EMU) in 2001. This development allowed Greeks to borrow and consume more *and* delayed the market's realisation of the Greek sovereign debt problem. At the same time, the EU's monitoring mechanisms failed. These failures put things into perspective; nonetheless, they are not sufficient reason to hold fellow EU members responsible for Greece's present economic predicament. The Greek economic policies of the last three decades have brought Greece close to bankruptcy. While Greeks kept on spending, the Greek state proved incapable of collecting revenues efficiently.[2] On top of that, the dysfunctional Greek judicial system has not helped to combat the prevalent culture of impunity (Tsakyrakis, 2010).

Finally, another crucial proximate cause of the Greek crisis has been the inaction of Greek political elites. Not surprisingly, the two main parties in Greece have been pushing the envelope for a while. In the most recent cycle of blame, the Karamanlis government attributed the increased 2004 deficit to the cost overruns of the 2004 Olympic Games and on the concealment of the true picture of Greek public finances by PASOK (the Panhellenic Socialist Movement) (BBC, 2004). A financial audit was conducted and the centre-right government promised transparency from that point forward (Eurostat, 2004). However, the Karamanlis government did not manage to implement the necessary structural reforms during its five years of rule. Moreover, the Karamanlis government did not succeed in addressing the problem of rapid debt accumulation and uncontrolled fiscal deficit. In September 2009, Kostas Karamanlis called a snap election under the pressure of a rising Greek deficit, a global financial crisis and vociferous opposition (Kyriakidou, 2009). At the apogee of the financial crisis, the then-governing centre-right party (Nea Demokratia) lost the election to PASOK. As soon as it came to power, PASOK revealed a far bleaker picture of the Greek economy than anyone had expected – or purported to expect. Once again, the new government blamed the previous government and promised transparency.

This bleak state of the Greek economy was initially treated and perceived primarily as a challenging moment in EU–Greece relations. Greece's credibility as an EU member and partner had been tarnished. In the meantime, the new government, which had been elected on a "spending" rather than an "austerity" platform, did not take any of the necessary measures to decrease the deficit during the first few months of its rule. From October

[2] This situation reached its apex under the latest Karamanlis government.

2009 to March 2010, the newly elected PASOK government refused to accept the situation. In the meanwhile, Nea Demokratia was internally divided and focused on changing its leadership. During this period, Greece was especially vulnerable to speculative attacks by financial institutions that bet on Greece going bankrupt.

Moreover, Eurozone countries, and the EU as a whole, took longer than they should have to address this problem. As a result the euro plunged. Only after Christmas 2009, when the euro started to drop relative to the dollar, was the public debt crisis seen as a European problem. But soon after it was understood as a global problem, a sovereign debt problem affecting governments across the world, including Ireland, Spain, Portugal, Japan and even the US. Thus, the global financial crisis proved to be the catalyst that revealed Greece as the weakest link in the Eurozone.

The Socio-economic Consequences

Current public discourse is dominated by the financial crisis. Having engaged in conversations about the crisis with people as varied as taxi drivers, businessmen, politicians, young professionals, artists, unemployed people and public sector employees, I have noted four distinct attitudes towards the crisis. The first attitude is one of pessimism, whining and cynicism. The second consists of long diatribes about which way to go: Should Greeks endure the austerity measures of the joint IMF-EU bailout or should they go at it alone by restructuring the debt and returning to the drachma? A third group focuses on the repercussions that the financial crisis has had on their income and their future prospects. Fourth, and last, some focus on the scandals and punishment of the politicians who are identified as most responsible. Regardless of the mode of approaching the financial crisis, the everyday Greek suffers from its socio-economic consequences.

Negative Consequences

With the passage of time the economic consequences of the crisis have crystallised: a gross external debt of 185% of GDP, a negative real GDP growth rate (around -4%), increased taxation, inflation, shrinking household incomes and declining consumption. In the private sector, many businesses are closing, others are moving to Cyprus or Bulgaria (where taxation for businesses is at 10%) and large sums of capital have moved out of Greek banks. High taxation and inflexible labour laws have forced many companies to relocate and have almost eliminated foreign direct investment.[3]

The persistent problem with government revenue collection has been exacerbated and the economy is threatened with an even deeper recession. The middle class – the backbone of democratic politics – is emaciated. Small and medium-sized businesses are closing one after the other, more and more cheques bounce and non-performing bank loans are on the rise. Moreover, employees are losing their jobs (12% unemployment) and a new class is being formed, that of the "former middle class" (Pagoulatos, 2010).

All these developments have had a negative impact on the self-esteem of many Greeks. The dire economic situation has led many to desperate acts such as committing suicide. According to an NGO that monitors suicides in Greece for the Ministry of Health, "suicides have doubled, if not tripled" compared to the number in 2009 (*Kathimerini*, 2010b). According to the psychologist Aris Violatsis, "men who are no longer earning enough money to provide for their families and feel they no longer have a role to play – people who are going through an identity crisis" are more at risk.

Many young people have decided to leave the country and look for better opportunities abroad. Greece's greatest asset is its human capital. Brain drain has been a problem for Greece and this crisis is exacerbating it (Kitsantonis, 2010). Seven out of ten people who have just completed their studies (or are about to complete them) would gladly leave Greece for a job opportunity abroad. Half of them have already tried to get such a job (Hiotis, 2010b). People who previously would not have left are leaving, and the people that might have been planning to return are less likely to do so now. Sadly, another generation will face blocked social and political mobility as a result of nepotism, clientelism and corruption.

The delegitimisation of political parties and state institutions in Greece is a highly disconcerting development. In a recent poll only 39% of people said they would vote for PASOK or Nea Demokratia. This is a historic low for Greece (Hiotis, 2010a). Furthermore, eight out of ten citizens express

---

[3] For example, Coca-Cola 3E and perhaps TITAN are on their way out (Petridis, 2010).

disappointment with the government, and nine out of ten express disappointment with the main opposition party (*Kathimerini*, 2010a). The financial crisis, the numerous political scandals and the solutions proposed by the ruling party – in line with the IMF-EU recommendations – have led many people to believe that the social contract is up for renegotiation. An unfortunate consequence arising from the perception that government has let down the people is that it grants legitimacy to the use of violence in politics among certain individuals in Greek society.

The financial crisis has multiplied and exacerbated belief in conspiracy theories. For example, such theories proliferated after the Marfin Bank arson in early May 2010. Some argued that it was anarchists who wanted to punish the employees who were not on strike. Others argued that this was government-planned provocation aimed at delegitimising public unrest and the reaction to government policies. Many suggested that this was an outside job. In particular, according to this narrative, certain financial institutions and investors have bet on a Greek default and that is why they arranged this incident.

There are many more stories. What we are really dealing with is a daily phenomenon. To be sure, this phenomenon is not new or unique to Greece. Many societies resort to such explanations to satisfy the psychological needs of the people, to attribute blame to whomever they do not trust and to make sense of the world. However, the phenomenon has reached new levels this time around, both in intensity and in complexity.

Increasingly the conspiracy theories circulating in Greece have ceased to be plausible, while also multiplying in number. Thus, while in the past an event would usually have two interpretations, the official and the conspiratorial, nowadays it has three or four, if not more, competing explanations. The financial crisis has evolved into a real economic crisis, expressing itself in job loss and diminished purchasing power for the average Greek. This has led the population to new levels of uncertainty about the future, which provides a fertile ground for conspiracy theories to emerge and spread. Add to the mix the advanced technological capabilities – as compared to the past – that are available to most Greeks and the seriousness of the situation becomes apparent. An unwanted consequence of our ability to communicate freely and instantly is the dissemination of rumours and conspiracy theories that reproduce a culture of avoiding responsibility.

Tertiary education is at a historic low and the sovereign debt crisis is not helping. The cuts in allowances and bonuses have seriously affected university professors and especially lecturers and non-tenured assistant professors. Lecturers especially, with a base salary similar to that of an elementary school teacher, find it difficult to continue their necessary research activities, since they are either forced to work a second job (so no time is left for research) or they have no money for research. Inevitably, this situation increases their dependence on the tenured professors who control their academic future. Under these circumstances, the future of non-tenured Greek professors is increasingly less related to their research and merit (to the extent that it ever was) and increasingly in the hands of a arbitrary clientelistic system revolving around departmental and dean elections. With basic gross salary of €1,183 at the rank of lecturer, incentives for research are few for low-ranking and non-tenured academics in Greek universities today, while their dependence on personal relationships that are unrelated to their research is growing.

Of course, even the salaries of tenured professors do not correspond to their studies and qualifications or compare to those abroad, when their net monthly earnings are less than €3,000. This leads some of them to extra-curricular activities unrelated to their teaching obligations and at times even to their research interests. In general, the competitiveness of the Greek university system, and therefore of its graduates, is at its lowest point in the last 30 years.

Neutral Consequences

There are some signs that migration to Greece has decreased while a significant number of existing immigrants are moving out of Greece. This development is having a mixed impact. On the one hand, the immigrants who are leaving are the ones who contributed most to the Greek economy (Koleka, 2010). On the other hand, the extra cost of policing and integration efforts burdens an already bankrupt state.

Half of Greece's population lives and works in Attica. Decentralisation has been a goal for a long time because the rest of the country's infrastructure and economic development has lagged behind. Many Greek governments have had decentralisation as a stated goal in their political programmes. What years of policy planning have been unable to achieve, the financial crisis may compel. People who moved from Thessaloniki to Athens, from Kilkis to Thessaloniki or from Herso to Kilkis might find it advantageous to return to their place of origin. This may boost local economies and – if followed by the right set of incentives – set the basis for sustainable decentralisation.

Positive Consequences

This crisis may show itself as an opportunity for fundamental changes. The delegitimisation of the current political system may lead to real political change. In addition, it may lead to the prosecution of those who have embezzled money from the state. Many of the reforms that the current government has pursued – under the pressure of the IMF and the EU – are necessary and long overdue, but more are needed. Through these necessary reforms, the Greek economy will become more competitive and attract foreign investments. As a result, serious investments in tourism, private education, shipping, research and development, renewable sources of energy and proper services may occur (Azariadis, Ioannides and Pissarides, 2010). Additionally, the need for funds to cover daily government operations – while it remains difficult for the government to borrow from the international markets – is fundamentally changing the tax collecting processes. This in turn will put pressure on the unregulated economy. But, most importantly, we might change our mentality.

Epilogue

Let me close with a discussion of the mentality problem I referred to above. Each time a scandal is revealed involving the public sector most television analyses focus on whether the accusations were valid or not and who was to blame. It is definitely important that hiring in the public sector be according to merit and impartial; however, I believe a deeper problem is revealed when we consider this phenomenon. Nobody comments on the fact that all of these highly connected and educated people want to work in the public sector. It does not cross anyone's mind to question that motivation; to wonder why Greek society has ended up turning the public sector into *the* dream of young people. The crowning of the public sector as king of opportunities is the most important message. Changing this mentality is the key to Greece's recovery.

Such a mentality has bloated the public debt through yearly high deficits. In order for growth rates to rise, entrepreneurship initiatives that would attract foreign direct investments are needed. Nevertheless, this can hardly happen when most of the Greek population's dream is to enter the public sector. This indeed is the outcome of decades of patronage politics, populism and nepotism; all causes of the current crisis as highlighted above. Shrinking the public sector while at the same time achieving positive growth rates is the only solution. However, nothing can change until our mentality does. We need a movement that will campaign for entrepreneurship and innovation instead.

The European Union may actually emerge stronger as a result of this crisis, replacing the current emergency fund with a permanent bailout and restructuring mechanism, a European Monetary Fund (Mylonas and Vreeland, 2010). Austerity is needed in multiple corners of the EU to balance public spending and tax revenues. All in all, Greece may be a unique case in that it is experiencing the crisis of a failed public sector, with the root causes of its failings being different from Ireland's crisis, which primarily involves the private banking sector. Yet if Greece undertakes the right measures and the mentality of the people is altered, Greece may not become a failed developed state after all.

References

Alogoskoufis, G. (1995). Greece. The two faces of Janus: Institutions, policy regimes and macroeconomic performance. *Economic Policy*, *10*(20), 149–84.

Azariadis, C., Ioannides, Y. M., and Pissarides, C. A. (2010). Development is the only solution: Seventeen proposals for a new development strategy. 6 October. Available at http://greekeconomistsforreform.com/wp-content/uploads/A-I-P-DEVELOPMENTw.-abs-10-06-10.pdf, accessed 24 November 2010.

BBC. (2004). Greek debt spirals after Olympics. 12 September. Available at http://news.bbc.co.uk/2/hi/business/3649268.stm, accessed 24 November 2010.

Eurostat. (2004). *Report by Eurostat on the revision of the Greek government deficit and debt figures*. 22 November. Available at http://epp.eurostat.ec.europa.eu/cache/ITY_PUBLIC/GREECE/EN/GREECE-EN.PDF, accessed 24 November 2010.

Hiotis, V. (2010a). Crash for the parties (in Greek). *To Vima*, 4 July. Available at http://www.tovima.gr/default.asp?pid=2&ct=32&artid=341308&dt=04/07/2010, accessed 24 November 2010.

Hiotis, V. (2010b). I want to leave Greece (in Greek). *To Vima*, 29 August. Available at http://www.tovima.gr/default.asp?pid=2&ct=1&artid=351032&dt=29/08/2010, accessed 24 November 2010.

Hürriyet Daily News (2010). Book claims Greek crisis stems from Ottoman rule. 4 October. Available at http://www.hurriyetdailynews.com/n.php?n=book-claims-greek-crisis-stems-from-ottoman-rule-2010-10-04, accessed 24 November 2010.

Kaplan, R. D. (2010). For Greece's economy, geography was destiny. *New York Times*, 24 April. Available at http://www.nytimes.com/2010/04/25/opinion/25kaplan.html, accessed 24 November 2010.

Kathimerini. (2010a). Disappointment of voters towards the political system grows (in Greek). 10 October. Available at http://news.kathimerini.gr/4dcgi/_w_articles_politics_2_10/10/2010_418257, accessed 24 November 2010.

Kathimerini. (2010b). Suicides have doubled (in Greek). 4 August. Available at http://news.kathimerini.gr/4dcgi/_w_articles_ell_2_04/08/2010_410313, accessed 24 November 2010.

Kitsantonis, N. (2010). Young Greeks seek options elsewhere. *New York Times*, 14 September. Available at http://www.nytimes.com/2010/09/15/business/global/15drachma.html, accessed 24 November 2010.

Koleka, B. (2010). Hard times in Greece prompt Albanians to return home. *Reuters*, 1 June. Available at http://www.reuters.com/article/idUSTRE6503WB20100601, accessed 24 November 2010.

Kyriakidou, D. (2009). Greek PM calls snap election, blames economic crisis. *Reuters*, 2 September. Available at http://www.reuters.com/article/idUSTRE5815ZZ20090902, accessed 24 November 2010.

Legge, T. (2010). Bailouts in Europe: A punitive Versailles or a benevolent Marshall Plan? The German Marshall Fund of the United States. 2 December. Available at http://blog.gmfus.org/2010/12/02/bailouts-in-europe-a-punitive-versailles-or-a-benevolent-marshall-plan/, accessed 6 December 2010.

Lyrintzis, C. (1987). The power of populism: The Greek case. *European Journal of Political Research*, *15*(6), 667–86.

Mavrogordatos, G. (1993). Civil society under populism. In R. Clogg (Ed.), *Greece, 1981–89: The populist decade*, 47–64. New York: St. Martin's Press.

Mavrogordatos, G. (1997). From traditional clientelism to machine politics: The impact of PASOK populism in Greece. *South European Society and Politics*, *2*(3), 1–26.

Meghir, C., Vayanos, D., and Vettas, N. (2010). The economic crisis in Greece: A time of reform and opportunity. 5 August. Available at http://greekeconomistsforreform.com/wp-content/uploads/Reform.pdf, accessed 24 November 2010.

Mylonas, H., and Vreeland, J. (2010). Does the Eurozone need its own Monetary Fund? *The National* (Abu Dhabi), 4 April.

Pagoulatos, G. (2010). The Greek economic crisis: What it means for Greece and for Europe. 24 March. Available at http://www.youtube.com/watch?v=NSyR8T7zSEw, accessed 24 November 2010.

Pappas, T., and Assimakopoulou, Z. (forthcoming). Political entrepreneurship in a party patronage democracy: Greece. In P. Kopecký and M. Spirova (Eds.), *Party patronage and party government: Public appointments and political control in European democracies*. Oxford: Oxford University Press.

Petridis, G. (2010). "Titan" is about to leave Greece (in Greek). *Proto Thema*, 24 October, 7.

Straubhaar, T. (2010). Europe's fate is staked to the euro. The German Marshall Fund of the United States. 2 December. Available at http://blog.gmfus.org/2010/12/02/europes-fate-is-staked-to-the-euro/, accessed 6 December 2010.

Tsakyrakis, S. (2010). Memorandum about justice (in Greek). *To Vima*, 8 August. Available at http://www.tovima.gr/default.asp?pid=2&ct=122&artId=347660&dt=08/08/2010, accessed 6 December 2010.

Four Waves of Financial Crises in 40 Years: The Story of a Dysfunctional International Monetary Arrangement*

Robert Z. Aliber

In the last 35 years, there have been four waves of financial crises, all of which have shared certain key features. Each has involved substantial declines in asset prices, often real estate. And all four have seen the failure of financial institutions in three or four or even more countries at the same time. Many banks have failed and have had to be rescued by their governments. In the absence of government financial assistance, the depositors in these banks would have lost much of their money.

The first wave began in mid-1982 and involved Mexico, Brazil, Argentina and about 10 other developing countries, including South Korea, Venezuela, Peru, Chile, Nigeria, Indonesia and Algeria. The second wave occurred in the early 1990s and involved Japan and three of the Nordic countries: Finland, Norway and Sweden – a very different combination. Banks failed, although most kept their brand names thanks to capital from their governments. The Japanese banks' loan losses were two to three times greater than their capital, but there was never a run on these banks because it was understood that the depositors had 100% loan guarantees. The Asian financial crisis was the third wave. It began in early July 1997 when the Thai baht depreciated sharply. The Malaysian ringgit immediately came under tremendous financial pressure, as did the Indonesian rupiah, which lost more than 80% of its value. The Philippines was caught up in this crisis; and then, in November and December, South Korea was subject to a massive run on its currency, which lost a third of its value in several weeks. Then Russia, Brazil and Argentina were affected: banks in all of

* This text is derived from a lecture Professor Aliber delivered at an event organised by the Konstantinos Karamanlis Institute for Democracy in Athens, on 14 May 2010.

these countries failed; Argentina defaulted on its external debt. The fourth wave of crises began with the failure of Lehman Brothers in the second week of September 2008. This came in the wake of the decline in US house prices that began in December 2006 and continued at a moderate rate in 2007 and the first few months of 2008. Many mortgage brokers went bankrupt as a result. Then several of the hedge funds managed by Bear Stearns failed. In February 2008 Bear Stearns was subject to two runs: one on its share price, which destroyed a tremendous amount of shareholder value, and the second on its IOUs as lenders would not renew their short-term loans to Bear Stearns. In September 2008 the US government took over the ownership of Fannie Mae and Freddie Mac, both large government-sponsored mortgage lenders. Iceland's three large banks collapsed at the same time that Lehman failed.

What were the antecedents of these failures? The Latin American crisis of 1982 came after a decade in which the external debts of this group of developing countries had risen at the rate of 20% a year and bank loans to these countries had increased by 30% a year.

Consider the pattern of cash flows between the borrowers and the lenders. It is as if the bank lenders had said, "We will allow you to increase your indebtedness at a rate of 20% annually, and we will charge you interest of 8% a year." Imagine that at the beginning of the first year, the borrower owes the bank $100. The bank increases its loans by $20, and the borrower pays $8 in interest to the bank – the borrower then has $12 in new cash to finance infrastructure investment, fiscal deficits or high consumption. At the end of the year, the borrower's debt has increased to $120. During the second year the borrower receives $24 in new cash and pays slightly more than $8 in interest. This is a marvellous world for the borrower because all the cash that is needed to pay the lenders comes from new loans.

This same pattern of cash flows was in evidence in the period leading up to the financial crisis that involved Japan and the three Nordic countries in the early 1990s. Property prices had been rising at a rate of 25–30% a year for at least five years. Stock prices had been increasing at comparable annual rates. Bank loans for the purchases of real estate had likewise been increasing at 25–30% a year. The same pattern of cash flows also occurred prior to the Asian financial crisis in each of the developing countries involved.

The term for this pattern of cash flows, when the rate of growth of the debt of a group of borrowers is two to three times the rate of interest, is

"Ponzi finance". In the early 1920s Charlie Ponzi operated a bank in the United States, and he promised to pay 45% interest every three months. He used the money obtained from selling new IOUs on Tuesday to pay the interest owed to those who had bought his IOUs on Monday.

A similar pattern of cash flows occurred in the United States after 2002. House prices increased at rates of 20–30% a year in 16 of the 50 states, and even more rapidly in some smaller markets. Mortgage interest rates were 5–6% a year. Those who bought houses made tremendous capital gains. They made small down payments, often no larger than 5–10%. Imagine that the house price increased from $100 to $120. If the buyer's down payment had been 10% of the purchase price, the rate of return on the buyer's cash investment would have been 200%. Millions of investors began to buy properties for these fantastic gains.

This pattern of cash flows is explosive and unsustainable. There is always an event – which can be called a "trigger" event– that leads the lenders to realise that they have extended too much credit to the borrowers. When the lenders curtail lending, some of the borrowers no longer have the cash to pay the interest on their outstanding loans, and they become distress sellers of assets.

In many of these episodes the borrowers went bankrupt because they had not seen that they were involved in an unsustainable pattern of cash flows; they could not adjust to the sudden decline in the availability of new cash. The situation in Greece today is similar to that of Mexico and Brazil in the early 1980s, with two significant differences. The first is that Mexico's and Brazil's debt-to-GDP ratios were lower than that of Greece. The second is that they had their own currencies. The failure of the borrowers to repay endangered the solvency of many of the lenders.

During the Asian financial crisis, some of the domestic banks failed because their domestic borrowers failed, and some failed because the depreciation of their currencies meant that the domestic equivalent of their IOUs, which were denominated in US dollars, surged in accordance with the decline in the value of their currencies.

The patterns of cash flows in the years prior to each of the four waves of crises were similar, even though, for the most part, different borrowers and lenders were involved. Some of the bank lenders have been involved in several of these waves. Thus the Japanese banks were involved in the first wave, the second wave and the third wave as well.

The financial crisis in Iceland is instructive, as are the years that preceded it. About 10–15% of the assets of Iceland's three large banks consisted of stocks in various Icelandic firms. The flow of foreign money into Iceland that began in about 2002 led to an appreciation of the Icelandic krona and to increases in asset prices in the country. The residents of Iceland who sold all or part of their securities to non-residents had to decide what to do with the money received from these sales. They had two choices: they could buy real estate and other securities, or they could buy consumption goods. To the extent that they bought stocks, stock prices increased, which meant that the value of bank assets, bank capital and household wealth also rose. The capital of the banks increased above the regulatory minimum. These banks then increased their lending to reduce their excess capital. Some of these loans went to individuals, including friends of the officers of the banks, who bought stocks. Bank earnings increased at a very rapid rate, primarily from the increases in the value of the stocks that they owned. Stock prices were rising by 50–60% a year, and bank capital was increasing at the same rate. The pattern was unsustainable: the day would come when stock prices would stop rising. When that happened, the earnings of the banks would become zero, but the price of bank stocks would be very high – for a brief while. Individuals would seek to sell bank stocks – but there would be no buyers. The market would be highly illiquid, and stock prices would plummet almost immediately. When Lehman failed in September 2008, the Icelandic banks collapsed. The value of the Icelandic currency fell by almost 50%.

Consider now the financial problems of the Club Med countries, which are similar. The ratio of their fiscal deficits to their GDPs is more than 10%. The ratios of government debt to GDP range from 70% to 125%. Their current account deficits are very large relative to their GDPs, and they have high levels of unemployment. These countries have five *nested* problems: each of these problems is nested within another.

The level of costs in each of these countries is too high relative to the level of costs in its trading partners in northern Europe. The high level of costs is the cause of the current account deficits. The large current account deficits explain why these countries have high levels of unemployment, which, in turn, is the cause of the large fiscal deficits. Because these countries have had large fiscal deficits for a number of years, the ratio of government debt to GDP is very high. But the underlying problem is that costs are too high – everything else is a symptom.

Consider the various policy measures that have been proposed: Are they directed at one or more of the symptoms or at the cause? The dominant policy measure is to reduce the fiscal deficit. Reducing the fiscal deficit will reduce costs modestly, but it will also reduce spending and lead to a higher level of unemployment.

What measures can these countries take to reduce their costs? One is that everyone voluntarily takes a 5–10% cut in salary; in this way the measures that have been applied to the government sector would be applied to the private sector as well. Deflation will come about automatically, but extremely slowly, as a result of high levels of unemployment. An alternative is that one or several of these countries take a holiday from the European Monetary Union. A holiday is like a separation. However, there are no provisions in the Maastricht Treaty for such a separation. Therefore, government officials in the countries that have large fiscal deficits should be thinking of economic scenarios in which it would be in the interest of their constituents to take a temporary holiday from the commitment to the common currency.

If governments have not been considering such a separation, then they have been derelict in their duty. But it is obvious that the governments have been derelict because none of these countries would have got into the current mess if their governments had asked: "Where are we going to get the money to finance our fiscal deficit when the foreign banks stop lending the money to us?"

The Future of Economic Governance in the European Union*

Christos Gortsos

Economic Governance in the European Union According to the Provisions of Primary and Secondary European Law

The Content of the European Economic Union

The content of the European economic union, the first of the two elements of the European Economic and Monetary Union (hereinafter the EMU), is defined in article 119, par. 1, of the Treaty on the Functioning of the European Union (the TFEU) (OJ, 2008, pp. 47–199), according to which, "for the purposes set out in article 3 of the Treaty on European Union, the activities of the member states and the Union shall include, as provided in the Treaties, the adoption of an economic policy which is based on the close coordination of member states' economic policies".[1]

Based on this definition, it can be concluded that, in contrast to the European monetary union element of the EMU, in the context of which member states that have adopted the single European currency, the euro,

---

* This article is based on the Appendix of a written submission prepared for and submitted to the UK House of Lords in October 2010.

[1] See also article 5, par. 1, of the TFEU. In addition, article 120 of the TFEU sets out the scope of the economic policy to be pursued by member states and the context within which this should be conducted in order to fulfill the objectives set by the TFEU. It is worth mentioning that the provisions of the TFEU on the European economic union (articles 120–6) reflect (with only marginal modifications of an institutional nature) those of the Treaty establishing the European Community (articles 98–104) in force since November 2003, which was superseded by the TFEU in 2009.

have abandoned their monetary sovereignty (including their power to conduct an autonomous monetary and foreign exchange policy) (OJ, 2008, pp. 47–199, articles 127, par. 2, and 219), the economic (mainly budgetary) policy of member states has not become European.[2] Accordingly, there is a stark asymmetry between the two elements of the EMU, that is, the monetary union and the economic union.

In this regard, with the beginning of the third stage of the EMU (on 1 January 1999), no member state of the European Union, no matter whether it has adopted the euro as a single currency or not, has surrendered its autonomy in the conduct of its budgetary policy. Fiscal policymaking remains decentralised.

However, this principle of budgetary autonomy has been framed by the provisions of the TFEU pertaining to the European economic union and referring to

- the principle of coordination of member states' economic policies; and
- the principle of fiscal discipline.

The Principle of Coordination of Member States' Economic Policies

Even though the conduct of budgetary policy (including both expenses and financing) and other economic policies (excluding monetary and foreign exchange policies, which have been "Europeanised" for the euro area member states) remain at the disposal of national governments, their coordination has been framed by the provisions of the TFEU. According to the principle of economic coordination, the economic policies of member states, while autonomous, must converge in order to create an environment of strong surveillance of member states' economic policies.

Article 121, par. 1, of the TFEU states in this respect that the member states should regard "their economic policies as a matter of common concern and … coordinate them within the Council, in accordance with the provisions of article 120".

The principle of coordination of member states' economic policies contains two components:

---

[2] The "Europeanisation" of member states' economic policies, if it were ever to be achieved, would lead to (and necessarily presuppose a political decision for) European political unification.

- the adoption by the Council of recommendations setting out broad guidelines for the economic policies of member states and the European Union;[3] and
- the procedure of multilateral surveillance of the economic policies of member states, conducted by the Council based on reports submitted by the European Commission (OJ, 2008, pp. 47–199, article 121, par. 3–4).

The latter provisions are further specified by those of Council Regulation 1466/97 on strengthening the surveillance of budgetary positions and the monitoring and coordination of economic policies (OJ, 1997, pp. 1–5),[4] as in force,[5] which is the first component of the so-called Stability and Growth Pact (SGP), which constitutes secondary European law.

The Principle of Fiscal Discipline

The principle of fiscal discipline is the third pillar pertaining to the functioning of the European economic union. According to this principle, member states are required to rationalise the methods of financing their public expenses and avoid excessive government deficits and public debt. This has been considered necessary in order to contribute to the main task assigned to the European System of Central Banks, consisting of maintaining price stability in the euro area (TFEU, article 127, par. 1).

The mechanism for imposing fiscal discipline contains, as well, two components:

- the imposition on member states of specific prohibitions with regard to the financing of their government deficits (TFEU, articles 123–5);[6] and
- the imposition of a detailed procedure (imposing fines in the case of non-compliance) with regard to excessive government deficits (TFEU, article 126).

---

3 It is worth mentioning that, due to their specific importance in shaping national economic policies, these recommendations were adopted by the Ecofin Council (TFEU, article 121, par. 2, third sentence) *after* a discussion, on conclusions, of the draft broad guidelines at the European Council (OJ, 2008, pp. 47–119, articles 127, par. 2, and 219; article 121, par. 2, second sentence).

4 OJ, 1997, pp. 1–5.

5 Amended by Council Regulation (EC) 1055/2005 (OJ, 2005, pp. 1–4).

6 See also the relevant provisions of Council Regulations 3603/93 (OJ, 1993, pp. 1–3) and 3604/93 (OJ, 1993, pp. 4–6).

These provisions are further specified by the following:

- Protocol (No. 12) to the Treaties on the excessive deficit procedure (OJ, 2008, pp. 279–80), and Protocol (No. 13) to the Treaties on the convergence criteria (OJ, 2008, pp. 281–2), which constitute primary European law;[7] as well as
- Council Regulation 1467/97 on speeding up and clarifying the implementation of the excessive deficit procedure (OJ, 1997, pp. 6–11), as in force,[8] which is the second component of the above-mentioned SGP.

Towards a Reinforcement of the Existing Economic Governance in the European Union

Intermediate Initiatives Taken as a Result of the Recent (2010) Euro Area Fiscal Crisis

Measures Taken in the Context of the Greek Debt Crisis

As a result of the recent (2007–9) international financial and economic crisis, the public finances of several member states were negatively affected for various reasons (some of which are common, others idiosyncratic), which in turn led in 2010 to a euro area fiscal crisis. The most severe case has been that of Greece, which has proved unable, during the first quarter of 2010, to refinance its debt on international bond markets.

In order to overcome the negative spillover effects from the potential need of the Greek government to resort to a rescheduling of its debt, the Eurogroup (the meeting of Ministers of Finance of the euro area member states), the European Council, the meeting of the heads of state or government of the euro area, and the President of the European Commission announced a combined euro area member states/IMF package for Greece of, initially, €30 billion (Statement, 11 November 2010) and then €120 billion.[9]

---

[7] See article 51 of the Treaty on European Union (OJ, 2008, pp. 13–45).

[8] Amended by Council Regulation (EC) 1056/2005 (OJ, 2005, pp. 5–9).

[9] Activated on the basis of findings of the Eurogroup (Statement by the Eurogroup, 2 May 2010).

The Creation of a European Financial Stabilisation Mechanism

In the face of a potential generalised debt crisis in the Eurozone (in the wake of the Greek debt crisis and the downgrading by rating agencies of the public debt of several other euro area member states), the Ecofin Council adopted a "European financial stabilisation mechanism" (Press Release 9596/10)[10] on 9 May 2010 in order to restore confidence on the international financial markets.

This mechanism consists of three elements:[11]

- First, an extension of the existing EU balance of payments finance facility for member states whose currency is not the euro[12] to euro area member states, based on article 122, par. 2, of the TFEU, for a total of €60 billion. The release of instalments of any loans under the €60 billion EU programme is to be decided upon by the Commission on the basis of compliance with the adjustment programme a member state submits and with the economic policy conditions adopted by the Council by qualified majority voting.
- Second, an intergovernmental agreement of euro area member states providing for loans to individual euro area member states, through a Special Purpose Vehicle (SPV) of the other euro area member states of up to €440 billion. The text of the legal instrument underpinning the European Stabilisation Mechanism is supposed to ensure that the SPV will borrow on the financial markets backed by guarantees of the euro area member states, limited to their respective national central banks' capital share in the European Central Bank. Loans of the SPV can only be granted unanimously by the euro area member states.
- Third, IMF participation equal to at least half of the former amounts (€250 billion).

---

[10] See also the Decision of the Representatives of the Governments of the Euro Area Member States Meeting within the Council of the European Union and the Decision of the Representatives of the Governments of the 27 EU Member States.

[11] See the Council Regulation (EU) 407/2010 of 11 May 2010 establishing a European financial stabilisation mechanism (OJ, 2010, pp. 1–4).

[12] See the Council Regulation (EC) 332/2002 of 18 February 2002 establishing a facility providing medium-term financial assistance for member states' balances of payments (OJ, 2002, pp. 1–3), last amended by Council Regulation (EC) 431/2009 of 18 May 2009 (OJ, 2009, pp. 1–2).

The fact that activation of the mechanism will take place in the context of joint euro area member states/IMF support is understood to ensure strict conditionality (by the IMF, according to its rules of operation). The legal texts seek to avoid difficulties with regard to the so-called no bail-out clause of the TFEU (article 125).

Recent (September 2010) Proposals of the European Commission

General Overview

The above-mentioned developments have underscored the need for stronger and more effective economic policy coordination, extending beyond budget deficit procedures. This was acknowledged by the European Council, which in March 2010 requested that its President, Herman Van Rompuy, explore ways to strengthen economic policy coordination. Taking into consideration the work undertaken, accordingly, by the Task Force on Economic Governance chaired by the President of the European Council, and based on its previous communications on economic governance dated 12 May and 30 June (IP/10/561 and IP/10/859), on 29 September 2010 the European Commission adopted a legislative package containing proposals for a comprehensive reinforcement of economic governance in the EU and the euro area. This legislative package contains

- proposals for three Regulations and one Council Directive dealing with fiscal issues, including a wide-ranging reform of the SGP; and
- proposals for two Regulations aiming at effectively preventing and correcting emerging macroeconomic imbalances within the EU and the euro area.

All these proposals are compatible with the existing provisions of the TFEU. Accordingly, the Commission does not propose any revision of the primary, but only an enhancement of the secondary European law pertaining to economic governance in the EU.

Legal Acts Dealing with Fiscal Issues

Regulation amending the legislative underpinning of the preventive part of the Stability and Growth Pact (amendment of Council Regulation

1466/97).[13] According to the provisions of this proposal for a Regulation, EU member states will have to adopt prudent fiscal policies in good times (during economic growth) in order to build up the necessary buffer for bad times (during a recession). The monitoring of public finances will be based on the new concept of prudent fiscal policymaking, which should ensure convergence towards the medium-term objective.

For member states not at this objective, annual expenditure growth should be set below trend growth, unless extra revenues can be collected, in order to ensure convergence towards the objective. The Ecofin Council will be competent for ensuring that member states comply with the objectives of prudent fiscal policymaking. For euro area member states, the Commission may issue a warning in case of significant deviation from prudent fiscal policymaking.

Regulation amending the legislative underpinning of the corrective part of the Stability and Growth Pact (amendment of Council Regulation 1467/97).[14] According to the provisions of this proposal for a Regulation, debt developments will be followed more closely and put on an equal footing with deficit developments as regards decisions linked to the excessive deficit procedure (article 126, TFEU). Accordingly, the decision to open an excessive deficit procedure will be based on a wider range of criteria.

Member states will be benchmarked as to whether they can sufficiently reduce their debt. Those whose debt exceeds 60% of their GDP should take steps to reduce it at a satisfactory pace, defined as a reduction of 1/20th of the difference from the 60% threshold over the last three years.

Regulation on the effective enforcement of budgetary surveillance in the euro area.[15] The above-mentioned proposed modifications of the SGP are backed up, through the provisions of this proposal for a Regulation, by a new set of gradual financial sanctions for euro area member states. In particular,

- as to the preventive part, an interest-bearing deposit should be the consequence of significant deviations from prudent fiscal policymaking;

---

13 See COM, 2010c.

14 See COM, 2010b.

15 See COM, 2010e.

- as to the corrective part, a non–interest bearing deposit, amounting to 0.2% of GDP, should apply upon a decision to place a member state in excessive deficit; this would be converted into a fine in the event of non-compliance with the recommendation to correct the excessive deficit.

Accordingly, the SGP will become more rule-based, and sanctions will be the normal consequence for euro area member states breaching their commitments.

To ensure enforcement, a "reverse voting mechanism" is envisaged when imposing these sanctions. Accordingly, the Commission's proposal for a sanction will be considered adopted unless the Council overturns it by qualified majority.

Directive on requirements for the budgetary frameworks of member states.[16] Since fiscal policymaking is decentralised in the context of the EU economic governance in force, the objectives of the SGP should be reflected in member states' budgetary frameworks. The proposal for this Directive sets out minimum requirements to be followed by member states in order to ensure that these frameworks be strengthened and fully aligned with the new European economic governance rules, by

- ensuring consistent accounting systems;
- aligning national fiscal rules with the provisions of the TFEU;
- switching to multi-annual budgetary planning; and
- ensuring that the system of public finances is covered by the framework.

Regulations Aiming at Efficiently Preventing and Correcting Emerging Macroeconomic Imbalances Within the EU and the Euro Area

Regulation on the prevention and correction of macroeconomic imbalances.[17] This proposal for a Regulation introduces a new element in the EU's economic surveillance framework: the excessive imbalance procedure (EIP). This procedure will comprise a regular assessment of the risks of imbalances in a member state based on a scoreboard composed of economic indicators.

According to the provisions of this proposal,

---

16 See COM, 2010a.

17 See COM, 2010f.

- once an alert has been triggered for a member state, the Commission will launch a country-specific, in-depth review to identify the underlying problems and submit recommendations to the Council on how to deal with the imbalances;
- for member states with severe imbalances or imbalances that put at risk the functioning of the EMU, the Council may enact the EIP and place the member state in an "excessive imbalances position";
- a member state under EIP will have to present a corrective action plan to the Council, which will set deadlines for appropriate action; and
- repeated failure to take corrective action will expose the concerned euro area member state to sanctions.

Regulation on enforcement measures to correct excessive macroeconomic imbalances in the euro area.[18] As in the fiscal field (see above), if a euro area member state repeatedly fails to act on Council EIP recommendations to address excessive imbalances, it will have to pay, according to the provisions of this Regulation, a yearly fine equal to 0.1% of its GDP. The fine can only be stopped by a qualified majority vote (according to the reverse voting mechanism described above), with only euro area member states having the right to vote.

Conclusions

The recent (current?) fiscal crisis in the European Union, and in particular in the euro area, has proven that the design of the institutional framework pertaining to the European economic governance was, under conditions of stress, inefficient and insufficient. This is mainly a result of the fact that this framework was shaped in the early 1990s, a time when the possibility had not been seriously considered that

- a member state would have to resort to extraordinary measures to support the stability of the financial system during a recession; and
- a member state would be unable to service its public debt.

It should be considered in this context that the shaping of the European economic union was designed without any previous international relevant

[18] See COM, 2010d.

evidence. The forming of the EMU is a historically unique event and its economic element has been established under the political conditions prevailing during the period of its adoption, mainly in order to support, at a minimum, the proper functioning of the monetary element.

As a matter of fact, the recent (2007–9) international financial crisis has induced member states to take measures unprecedented for Europe since the 1930s. In several member states the financial system was close to a collapse, due both to the interconnectedness of insolvent credit institutions with the rest of the financial system and the real sector of the economy, as well as the fear of "banking panics". In addition, the severe recession in almost all member states has negatively affected the ability of their governments to perform efficiently in implementing their budgetary plans.

According to the provisions of the institutional framework in force pertaining to European economic governance before the occurrence of the financial crisis (see above), member states were bound by the principles of coordination of their economic policies and of fiscal discipline, while enjoying, in principle, autonomy with regard to the conduct of their budgetary policy. However, in the middle of the international financial crisis several member states were forced to take fiscal measures that induced them to deviate drastically from their budgetary plans. Hence fiscal discipline has, in several cases, been abandoned.

To the extent that fiscal policymaking remains decentralised, the responsibility to take measures in order to bail out parts of the financial system remains at the discretion of member states. Accordingly, under stress conditions, member states will necessarily have to deviate from sound fiscal practices, unless a new *efficient* institutional framework is put in place, which would

- provide for the orderly resolution of (insolvent) systemically important financial institutions without the intervention of national governments; and
- not impose excessive burdens on the financial system to the detriment of economic development (due to the higher interest rates that would have to be imposed by credit institutions on the financing of households and enterprises in order to compensate for the levies imposed on them to finance bank-based resolution projects).

It is worth mentioning that the institutional framework pertaining to European economic governance did not contain provisions on the management and resolution of fiscal crises in the EU and the Eurozone. It was only during the spring of 2010 that, for the first time, European institutions took emergency measures in this respect (see above).

Nevertheless, in my view, it is imperative that member states comply with sound fiscal rules, set at European level, especially during periods of economic growth. In this respect, any derogations from member states' compliance with the provisions requiring provision of adequate statistics and with the rules of multilateral surveillance should be addressed for correction. In this context, however, it is questionable whether the imposition of fines on member states which, *in extremis*, are unable to service their public debt is an efficient solution. Accordingly, effective prevention of fiscal crises is, in my consideration, definitively the best first choice within the framework of the European economic governance.

Strengthening the provisions of the SGP may be a solution to problems arising from member states' unwillingness or inability to comply with the institutional framework in force pertaining to European economic governance (see above). However, the limits to the effectiveness of these provisions (especially those referring to the corrective part) have become manifest during the recent crisis, as shown above.

Finally, there is no doubt that macroeconomic and competitiveness imbalances among member states should be reduced. Nevertheless, this issue pertains to structural reforms in the European economy that go beyond the shaping of economic governance within the context of the European economic union. A longstanding solution to this problem is a real challenge for the decades to come and requires the implementation of policies which, until now, have met with severe resistance from several member states.

References

COM (Commission of the European Communities). (2010a). *Proposal for a Council Directive on requirements for budgetary frameworks of the Member States*. COM(2010) 523 final. Brussels, 29 August.

COM (Commission of the European Communities). (2010b). *Proposal for a Council Regulation amending Regulation (EC) 1467/97 on speeding up and clarifying the implementation of the excessive deficit procedure*. COM(2010) 522 final. Brussels, 29 August.

COM (Commission of the European Communities). (2010c). *Proposal for a Regulation of the European Parliament and of the Council amending Regulation*

(EC) 1466/97 on the strengthening of the surveillance of budgetary positions and the surveillance and coordination of economic policies. COM(2010) 526 final. Brussels, 29 August.

COM (Commission of the European Communities). (2010d). *Proposal for a Regulation of the European Parliament and of the Council on enforcement measures to correct excessive macroeconomic imbalances in the euro area*. COM(2010) 525 final. Brussels, 29 August.

COM (Commission of the European Communities). (2010e). *Proposal for a Regulation of the European Parliament and of the Council on the effective enforcement of budgetary surveillance in the euro area*. COM(2010) 524 final. Brussels, 29 August.

COM (Commission of the European Communities). (2010f). *Proposal for a Regulation of the European Parliament and of the Council on the prevention and correction of macroeconomic imbalances*. COM(2010) 527 final. Brussels, 29 August.

Decision of the Representatives of the Governments of the Euro Area Member States Meeting within the Council of the European Union and a Decision of the Representatives of the Governments of the 27 EU Member States. Available at http://www.consilium.europa.eu/showFocus.aspx?id=1&focusid=478&lang=EN, accessed 18 October 2010.

IP/10/561. http://europa.eu/rapid/pressReleasesAction.do?reference=IP/10/561&format=HTML&aged=0&language=FR&guiLanguage=fr, accessed 18 October 2010.

IP/10/859. http://europa.eu/rapid/pressReleasesAction.do?reference=IP/10/859&format=HTML&aged=0&language=FR&guiLanguage=fr, accessed 18 October 2010.

OJ (*Official Journal of the European Communities*). (1993). L 332, 31 December.

OJ (*Official Journal of the European Communities*). (1997). L 209, 2 August.

OJ (*Official Journal of the European Communities*). (2002). L 53, 23 February.

OJ (*Official Journal of the European Union*). (2005). L 174, 7 July.

OJ (*Official Journal of the European Union*). (2008). C 115, 9 May.

OJ (*Official Journal of the European Union*). (2009). L 128, 27 May.

OJ (*Official Journal of the European Union*). (2010). L 118, 12 May.

Press release 9596/10 (Presse 108), 9/10 May 2010. Available at http://www.consilium.europa.eu/uedocs/cms_data/docs/pressdata/en/ecofin/114324.pdf, accessed 18 October 2010.

Statement by the Eurogroup, 2 May 2010. Available at http://consilium.europa.eu/uedocs/cmsUpload/100502%20Eurogroup_statement.pdf, accessed 18 October 2010.

Statement on the Support to Greece by euro area member states, 11 April 2010. Available at http://consilium.europa.eu/uedocs/cms_data/docs/pressdata/en/ec/113686.pdf, accessed 18 October 2010.

Selected Secondary Sources

Andenas, M., and Hadjiemmanuil, C. (1997). Banking supervision, the internal market and European monetary union. In M. Andenas, L. Gormley, C. Hadjiemmanuil and I. Harden (Eds.), *European Economic and Monetary Union – The institutional framework*, 375–417. International Banking, Finance and Economic Law Series. United Kingdom: Kluwer Law International.

Claessens, S., Herring, R. J., and Schoenmaker, D. (2010). *A safer world financial system: Improving the resolution of systemic institutions*. Geneva Reports on the World Economy, no. 12. International Center for Monetary and Banking Studies and Centre for Economic Policy Research.

Dewatripont, M., Rochet, J.-C., and Tirole, J. (2010). *Balancing the banks: Global lessons from the financial crisis*. Princeton: Princeton University Press.

Gortsos, C. V. (2006). Commentary of the Treaty on European Union and of the Treaty Establishing the European Community, Economic and Monetary Union (art. 98–124) (in Greek). In C. V. Gortsos, G. D. Kremlis, A. I. Konstandinides et al., *Commentary of the Treaty on European Union Treaty and of the Treaty Establishing the European Community*, vol. 3, *Treaty Establishing the European Community (art. 81–188)*, 127–282. Athens and Komotini: Ant. N. Sakkoulas Publishers.

Häde, U. (1999). Kommentar zu den Artikeln 4, 8 und 98–124 des EGV. In Ch. Caliess and M. Ruffert (Eds.), *Kommentar zu EU-Vertrag und EG-Vertrag*, 293–305, 362–3, 1101–1252. Neuwied: Luchterhand Literaturverlag, Perlentaucher.

Lastra, R. M. (2006). *Legal foundations of international monetary stability*. Oxford and New York: Oxford University Press.

Lastra, R. M., Louis, J.-V., Gortsos, C. V. et al. (2010, forthcoming). *Developments in European banking law*. Report of the Committee of the International Monetary Law of the International Law of Association (MOCOMILA) at the ILA biannual conference in The Hague.

Louis, J.-V. (2009). *L'Union européenne et sa monnaie*. 3rd ed. Brussels: Institut d'Études Européennes, Éditions de l'Université de Bruxelles.

Smits, R. (1997). *The European Central Bank – Institutional aspects*. The Hague: Kluwer Law International.

Entrepreneurship and Economic Development: The Changing Role of Government*

E. S. Savas

Over the past century, governments throughout the world have grown in size, in the amount they spend and in power. But people have come to expect too much from their governments – the governments that they enlarged, sustained and empowered with their votes and their taxes.

In recent decades people have begun to realise that governments are limited in what they can accomplish, that governments are unable to satisfy every expectation and to deliver on all their promises. As a result, the role of the state is changing at national, provincial and local levels of government.

The change is strategic. It is based on the realisation that government is only one of the four institutions created by human societies over thousands of years to provide the goods and services that people want and need. The other three institutions are the market, civil society and the family. Given the newly recognised limitations of traditional government and the huge economic and social problems we face, we must harness the unique abilities and strengths of all four of these institutions – government, the market, civil society and the family – to address these problems. Instead of continuing to rely excessively on overburdened governments, we should reallocate societal tasks to the institutions best able to handle them.

This reallocation calls for five policies that change the role of government. Together they create new areas for economic growth and development and new opportunities for entrepreneurs. The five policies are (1) government reverting to its core functions; (2) government adopting market prin-

---

* This text is derived from a lecture Professor E. S. Savas delivered at an event organised by the Konstantinos Karamanlis Institute for Democracy in Athens, on 13 October 2010.

ciples for its programmes; (3) government decentralising and devolving power and revenue sources to lower levels of government; (4) government relying more on civil society and the family; and (5) managing government performance and focusing on results, not merely spending. The result would be a sleek, limited, more nimble government better able to handle its true responsibilities.

Government Reverting to Its Core Functions

Let me start with the first policy, government reverting to its core functions. It has been said that "Government wastes more money doing the wrong things well than doing the right things poorly."[1] Therefore the first question to ask about a government activity is, "Should government be doing this?" Governments engage in many activities that are not among their core functions and can safely be left to the private sector – to companies or to non-governmental organisations (NGOs). That is, markets or civil society can take over and supply those goods and services that are not inherently governmental, thereby allowing government to shed a load that it can no longer bear.

There are numerous – and even humorous – examples of truly questionable government activities. For example:

- Pakistan International Airlines, which is state owned, has a poultry farm where it raises chickens (Ingram, 2010). Why is an airline also in the business of raising chickens? Why, to feed its passengers, of course! Even Olympic Air had better ways to provide in-flight meals.
- In Egypt, the state owned a brewery and made alcoholic beer – in a Muslim country! It also made Coca-Cola, manufactured automobile tires and had a factory that baked cakes and cookies.

But let me not pick examples from only these countries.

- The United States Navy owned a 350-hectare dairy farm that supplied the milk for the nearby Naval Academy, where 4,000 men and women train to become naval officers. Why did the US Navy own cows and a farm? Was it experimenting with a secret biological weapon? The explanation: In 1911 an outbreak of typhoid fever was traced back to con-

[1] This is often attributed, without authentication, to the Nobel Prize–winning economist Milton Friedman.

taminated milk. Therefore the Navy started its own farm to assure a safe supply of milk (Boothy443, 2007). Eighty-seven years later it was still supplying the milk for the Academy. But a study found that the total cost was 50 cents a gallon more than the retail price of milk in the local supermarket, and the farm was privatised.
- The US government took over a bankrupt freight railroad and spent $7 billion over 11 years trying to improve it, but when President Reagan finally sold it through a public sale of shares, it brought only $1.65 billion (Rand Herron and Miles, 1987). Under private ownership, however, the railroad thrived, expanded and employed many more workers, and in just 10 years it was worth $10 billion.
- Two more examples from the US: New York State owns golf courses and a ski resort. New York City owned radio and television stations; Mayor Rudolph Giuliani sold them for $20 million (Haberman, 1996a; Haberman, 1996b) and $207 million (Toy, 1995), respectively.

None of these is an inherently governmental function.

Many state-owned businesses involve more common activities. Electric utilities, telecommunications, transportation, energy, banking – all are necessary, all require *appropriate* regulation, but *none* requires government ownership or operation. What *is* required for rapid economic development is smaller government and less government; that is, one that consumes less of the nation's wealth and interferes less in the national economy. Government ownership in these sectors generally leads to poor investments, excessive hiring of relatives, friends and political supporters, and bloated government. In short, governments should stick to their core functions.

Besides non-core functions, there is another whole category of "wrong things" that governments do: excessive and inappropriate regulation of occupations and people's daily work. For example, in the US, regulation of the trucking industry was heavily biased against small operators. But after trucking was extensively deregulated in the 1970s, small truckers were able to enter and compete in the long-distance trucking business. Many new trucking firms were started and small firms grew. The cost of transporting goods dropped and so did the prices of those goods. Small, local manufacturers were able to ship their goods economically to more distant customers. They grew. More jobs were created. The whole country benefited (Gale Moore, 2002).

The same thing happened when airlines were significantly deregulated in the US. New airlines appeared, efficient routes were established, fares dropped and many more people started flying; air travel was democratised (Smith and Cox, 2008).

Years of state regulation produce bureaucratic obstacles and economic stagnation. For example, in Peru it took 289 days to navigate the labyrinth to register an industrial enterprise, longer than it takes to have a baby. And it took 26 months to get a license to operate a taxi (de Soto, 1989, p. 134) – something that takes one day in Hong Kong.

When economic activity is overregulated, regulators are able to suffocate businesses and extract bribes and thereby unwittingly throttle the economy. This is one of the reasons why India, the largest democracy in the world in terms of population, is currently lagging behind China in its pace of economic development.

Nations should abandon any economic model that resembles the one that failed in the Soviet Union. They should dismantle barriers to economic activity and allow more people the freedom to work, the freedom to enter occupations for which they are qualified and the freedom to start legitimate businesses and make them grow in a freer market.

The former Labour Prime Minister of Great Britain, Tony Blair, put it well when he said, "The only way we [the Labour Party can] win is by being the party of empowerment, and that requires a state that is more minimalist and strategic, that is about enabling people, about developing their potential but not constraining their ambition, their innovation, their creativity" (Stevens, 2010).

Government Adopting Market Principles for Its Programmes

Governments are adopting market principles for their programmes through several methods: privatisation, denationalisation, outsourcing, concessions, voucher systems, Public–Private Partnerships and user charges – at all levels of government. These methods open the way for entrepreneurs to build competitive enterprises by providing public services, while at the same time this policy reduces government spending. This is a double spur that helps revitalise a stagnant economy.

The first target for privatisation in some countries is the private sector itself, if it has long enjoyed a cosy, non-competitive symbiosis with the government. Many domestic companies are quite happy with the status quo: they sell their products and services to the state at high prices and buy goods and services from the state at low prices. The last thing they want is more private competitors and privatised state enterprises, because they would lose their favoured status and influence.

State-owned enterprises should be sold, given away or liquidated depending on their condition and their prospects. Besides the Western democracies, China, Vietnam and now Cuba have moved or are moving in this direction. International studies show that after denationalisation, surviving companies may grow, hire more workers, pay higher wages and pay more taxes (Savas, 2000, pp. 167–72).

Chile's experience supports this finding. Five state-owned companies in that country had a total of 12,500 workers. The companies were privatised, and six years later those same five companies employed 19,700 workers – increasing their workforce by more than 50%.

A different example from Chile (not involving one of those five companies) shows how the problem of overstaffing was managed. The state-owned telecommunications company was privatised and efficiency increased so much that only half as many workers were needed: the number of employees for each one thousand telephone lines, a standard measure of productivity, declined by more than half within four years, from 13.7 workers per thousand lines to 6.2. But the country had been suffering from inadequate telephone service. Many more telephone lines were needed, and the sale contract called for the buyer to double the number of lines within four years. As a result, no employees had to be discharged: they were retained, retrained and then managed properly in keeping with private-sector standards.

Two more examples further illustrate the great breadth of the privatisation movement:

- In the US, some public hospitals have been sold to non-profit and to for-profit companies. This has led to lower costs, better health care and greater patient satisfaction (Sataline, 2010).
- For the first 50 years of space travel, government involvement was necessary because of the huge capital requirements and extreme risks of this new and developing field. Now the space programme in the US is

being privatised: you can ride into space on a private spacecraft launched by a private company.

Privatisation is not limited to state-owned enterprises. It is also carried out at provincial and municipal levels by outsourcing local functions to private contractors under transparent, competitive conditions. This has been done with hundreds of common services, including bus transportation, water supply, waste-water treatment, road maintenance, street cleaning, garbage collection, park management, vehicle maintenance and many internal clerical functions, for example. Moreover, many social services are routinely contracted out to NGOs as well as to for-profit firms (Savas, 1987, pp. 73–4; Savas and Kondylis, 1993).

Let me give some examples of privatised public services to show the broad applicability of the concept:

- The Metro system in Stockholm, Sweden, is operated by a private French contractor.
- Parking meters on city streets have been privatised through concession arrangements. Chicago sold a 75-year concession on its 36,000 downtown parking meters for more than a billion dollars up front. Pittsburgh sold a 50-year concession for almost half a billion dollars. New York is considering doing the same for an estimated $5 billion, and Los Angeles is also looking into this option (Seifman, 2010; Dugan, 2010a).
- A number of cities in the US contract with a private company to operate their public libraries. All the policies of the libraries remain entirely under the control of the municipal governments, and the cities, not the library users, pay the contractor (Streitfeld, 2010b).
- In Denmark, two-thirds of the municipalities contract with a private company for fire and for emergency ambulance services (Falck, 2010). A majority of the population is protected through this arrangement.
- Several cities in the US have gone much further. In novel arrangements that have been in effect for more than five years, they contract out virtually all their municipal services to a private firm, keeping only the mayor, a city manager and police and fire departments. Almost everything else is done by the private company using its own employees. The largest city with this kind of public–private partnership is Sandy Springs, Georgia, with a population of more than 80,000 (Davis, 2010; Hunt, 2010). A California city with a population of about 40,000 did the same recently (Streitfeld, 2010a).

One of the best examples, and perhaps most useful immediately for Greece, is that of bus privatisation in Copenhagen, Denmark. The regional transportation authority controls a system that serves a population of more than two million people in the metropolitan area. It used to operate the buses as a public monopoly, but the Danish Parliament mandated a change to an entirely private contract system. The authority designs the bus routes and, through competitive bidding, hires private companies to operate different groups of routes under five-year contracts. Three foreign firms – one British, one French and one American – operate 81% of the routes and five small Danish firms operate the remainder. The costs were reduced by 22%, adjusted for inflation (Savas, 2002). The city of San Diego, California, privatised its bus service somewhat differently and reduced its costs by about 40% without reducing the level of service or its quality. Los Angeles, Las Vegas, Denver, Houston, Stockholm and London have had similar experiences (Savas, 2000, p. 152).

Numerous studies in many countries confirm that when outsourcing is done properly, costs are generally reduced by 20–30%, and by as much as 50%, while the level and quality of service are maintained or improved (Savas, 2000, pp. 147–55). Outsourcing creates openings for entrepreneurs to enter a new field – public services – and to deliver better public services at lower cost.

Besides divesting state enterprises and outsourcing government services, governments offer concessions for private groups to build public infrastructure (Dugan, 2010b; Merrick, 2010). Greece did this for the excellent new Athens airport and highway. These Public–Private Partnerships attract foreign and domestic private investment, create jobs and provide business opportunities and learning experiences for local companies. Airports, seaports, water systems, toll roads, bridges and dams are being built throughout the world by this method.

Brazil is planning to build up to 24 hydroelectric dams on the Amazon River and its tributaries using such Public–Private Partnerships. Foreign builders and investors will pay for the dams and then sell the resulting electricity over a period of decades, but the country will have power now to supercharge its growth into a world economic power. This is a marked change from the previous policy of hiring construction companies to simply build a dam and then turn it over to a state utility (Lyons, 2010).

Government Decentralising and Devolving Power

Let me refer briefly to the third policy I identified at the outset. Decentralising and delegating responsibility and authority – and assigning revenue sources – to local governments can lead to better, quicker, more cost-effective, more flexible and more innovative decisions and better public services. At the same time it nurtures the development of more local leaders – business entrepreneurs and social entrepreneurs – who are an asset in any nation and particularly in small ones, where people with such capabilities are few in absolute number. The new regional decentralisation in Greece should be a step in the right direction.

Government Empowering Civil Society and Families

The fourth policy is about empowering civil society. Civil society refers to the numerous informal, voluntary organisations based in communities and composed of people with similar interests. People are empowered to join together to address their problems directly as free citizens, instead of demanding government subsidies as their birthright or behaving as supplicants who are permanently dependent on government bureaucracies. Such organisations can be a powerful force for good: for example, in many social services and in community programmes for education, health care, recreation and job training (Goldsmith, 2010).

Managing Government Performance

The focus on better public management means setting objectives for government programmes, measuring the results in terms of efficiency and effectiveness and reporting the results to the public. This exposes problems and creates opportunities for innovative leaders. It reduces the cost and increases the productivity of government and leaves more resources for businessmen and women to invest in growing businesses and jobs. This requires a vigilant media and a concerned, not apathetic, public; the public has to pay attention to the performance and not only the politics of government.

These five policies are surfacing in governments throughout the world (Kettl, 2005). They unleash the abilities of enterprising individuals by creating niches and opportunities where their talents and fortunes can grow and benefit all – through communities, companies and NGOs. They lead to the proper goal of the state: a healthy society with a healthy economy.

Cultural and "Cratogenic" Problems

Let me close by referring to cultural and "cratogenic" problems. In Italy earlier this year I was teaching a doctoral class on privatisation. I was disappointed to learn that this group, consisting of some of the brightest young people in Italy, who are entering government service, consider it impossible to change government behaviour and futile to try. Most of them simply want easy, well-paying, lifelong jobs and early retirement with a fat pension. It was depressing to find them so passive. This is a looming cultural disaster. A few years ago I found the same attitude of fatalism and limited ambition among many students in Greece, but I am told that – fortunately – this is no longer the case.

I am proud – very proud – of my Greek heritage. But I confess to feeling embarrassed as I read about Greece's enormous problems and as I tried – defensively – to answer questions from my incredulous, non-Greek friends. Massive tax evasion, negligence in tax collection, jobs based on contacts rather than ability, swollen public payrolls, ridiculous pension arrangements and, throughout all, widespread and rampant corruption on an Olympian scale. This might be expected in some Third World countries, but surely not in the land that gave Western civilisation the concepts – and gave the English language the very words – "democracy", "govern", "politics" and "economics".

Greece is not alone in facing these problems. To the extent that such problems in any democratic nation have underlying cultural causes, the citizens deserve the governments they choose.

But often the problems may more accurately be characterised as "cratogenic", that is, given birth to or created by the state, and the state has perpetuated the problems for many years: subsidies granted as political favours; overstaffed public agencies whose complacent employees engage in make-work activities; workers trapped in sheltered cocoons doing obsolete jobs because of regulations that hinder labour mobility; people's citizenship skills atrophied from disuse; and the belief that the problem is not enough revenue, whereas the real problem, more likely, is too much spending. Such government failures tend to be prolonged and to get worse over time.

These government failures have to be fixed: less spending, greater productivity, the end of subsidies and the termination of unjustified projects. For example, in October 2010 the new governor of the State of New Jersey halted the construction of a $9 billion tunnel under the Hudson River to New York City because construction was started by the big-spending previous governor even though there was not enough money to build it and no plans to finance it (McGeehan, 2010). France is delaying the starting age for retirement pensions in order to reduce costs (Gauthier-Villars, 2010). Britain will slash 490,000 jobs from the public payroll over four years (CNN, 2010). In September 2010 alone, American cities and states eliminated 83,000 government jobs (*New York Times*, 2010). Cuba is laying off more than half a million government workers, 12% of its government workforce (de Cordoba and Casey, 2010; *Wall Street Journal*, 2010). In summary, governments in many countries, even in socialist Cuba, are being downsized, both in scale and in scope, and they are rediscovering the virtues of limited government and a market economy.

These transformations will be long, painful and heartbreaking. Each is a Herculean task that cannot be avoided if a nation is to survive. I remain confident that Greece can and will succeed.

References

Boothy443. (2007). US Naval Academy dairy farm (closed). Virtual Globetrotting, 14 April. Available at http://virtualglobetrotting.com/map/us-naval-academy-dairy-farm-closed/, accessed 29 October 2010.

CNN. (2010). UK budget cuts. wibc.com, 20 October. Available at http://www.wibw.com/nationalnews/headlines/UK_Budget_Cuts_105386863.html, accessed 29 October 2010.

de Cordoba, J., and Casey, N. (2010). Cuba to cut state workers in tilt toward free market. *Wall Street Journal*, 14 September, A1.

Davis, A. (2010). Is Georgia broke? Sandy Springs privatized. *MyFoxAtlanta,* 7 July.

Dugan, I. J. (2010a). Facing budget gaps, cities sell parking, airports, zoos. *Wall Street Journal*, 23 August, A1.

Dugan, I. J. (2010b). Private investors push into public projects. *Wall Street Journal*, 12 October, B1.

Falck. (2010). Falck emergency. Available at http://www.falck.com/businnes%20areas/Emergency/Pages/Emergency.aspx, accessed 29 October 2010.

Gale Moore, T. (2002). Trucking deregulation. In *The concise encyclopedia of economics*. Available at http://www.econlib.org/library/Enc1/TruckingDeregulation.html, accessed 29 October 2010.

Goldsmith, S. (2010). *The power of social innovation*. San Francisco: Jossey-Bass.

Gauthier-Villars, D. (2010). French pension bill clears big hurdle. *Wall Street Journal*, 28 October, A13.

Haberman, C. (1996a). Sell WNYC? Not so fast, critics say, as sale of TV station still awaits FCC approval. *New York Times*, 22 March, B3.

Haberman, C. (1996b). For WNYC Radio, there is a price for independence from an old master. *New York Times*, 5 July, B3.

Hunt, A. (2010). Sandy Springs takes fresh look at private services. *Atlanta Journal Constitution*, 15 March.

Ingram, F. C. (2010). Company history: Pakistan International Airlines Corporation. Answers.com. Available at http://www.answers.com/topic/pakistan-international-airlines, accessed 29 October 2010.

Kettl, D. F. (2005). *The global public management revolution* (2nd ed.). Washington, DC: Brookings Institution Press.

Lyons, J. (2010). Brazil engineers a critic-proof dam. *Wall Street Journal*, 7 October, A1.

McGeehan, P. (2010). Christie says he won't budge on scrapped rail tunnel. *New York Times*, 28 October, A25.

Merrick, A. (2010). Cash flows in water deals. *Wall Street Journal*, 12 August, A3.

New York Times. (2010). Public jobs drop and private hiring slows. 9 October, A1.

Rand Herron, C., and Miles, M. A. (1987). Sale of Conrail sets a record. *New York Times*, 29 March.

Sataline, S. (2010). Cash-poor governments ditching public hospitals. *Wall Street Journal*, 30 August, A3.

Savas, E. S. (1987). *Privatization: The key to better government*. Chatham, NJ: Chatham House Publishers.

Savas, E. S. (2000). *Privatization and public–private partnerships*. Washington, DC: CQ Press.

Savas, E. S. (2002). Interview with the deputy managing director of the HUR Transport Division, Copenhagen, 18 June.

Savas, E. S., and Kondylis, E. K. (1993). *Privatization and productivity* (in Greek). Athens: Papazisis Publications.

Seifman, D. (2010). City mulls $5B meter sell-off. *New York Post*, 4 October, 2.

Smith, F. L., Jr., and Cox, B. (2008). Airline deregulation. In *The concise encyclopedia of economics*. Available at http://www.econlib.org/library/Enc/AirlineDeregulation.html, accessed 29 October 2010.

de Soto, H. (1989). *The other path*. New York: Harper and Row.

Stevens, B. (2010). What Obama can learn from Tony Blair. *Wall Street Journal*, 5 October, A21.

Streitfeld, D. (2010a). A city outsources everything. California's sky doesn't fall. *New York Times*, 20 July, A1.

Streitfeld, D. (2010b). Outsourcing fairly healthy public libraries, town hears a roar. *New York Times*, 27 September, A1.

Toy, V. S. (1995). WNYC fans fear programming loss. *New York Times*, 13 August 1995.

Wall Street Journal. (2010). Cubans dip a toe in capitalist waters. 6 October, A17.

The Social Market Economy: A Cure for All Ills?

Anthony Ioannidis

The Social Market Economy

The Social Market Economy model is defined as an economic order which is mainly based on the free market but which includes elements of social balancing, namely, the principles of freedom, justice and solidarity (Hasse, Schneider and Weigelt, 2008).

The Social Market Economy was first documented during the economic miracle that took place in post-war Germany, and from then on it was seen as critical for ensuring economic prosperity and social justice. For many it holds the promise of "prosperity for all", while for some it offers the "cure for all ills".

According to idealists, the Social Market Economy will restructure political priorities by (a) ensuring lifelong access to further education for everyone; (b) creating social insurance systems; (c) addressing issues such as the protection of intellectual property; (d) giving developing countries fair access to free trade; and (e) putting into place new instruments and mechanisms that support voluntary self-responsibility (Eucken, 1982; Esping-Anderson, 1996; Esping-Anderson, 2000).

The Social Market Economy has to be seen as a privilege-free order, where neither feudal or party elites nor economic power groups like monopolies, cartels or trusts influence markets and society. Applied in principle, this translates into all members of a society receiving the same opportunities to develop individually, regardless of class. This results in improved welfare for everyone, which brings opportunities for consumption and a distribution of wealth within a society via the rule-based market order. Through the channels of mobility and redistribution of income over time by market forces, without governmental intervention, a socialisation of progress and profit takes place (Wrobel, 2010).

Within the European Union, four distinct clusters of welfare states can be identified (Esping-Anderson, 1990; John, 2007):

- The Nordic cluster (Denmark, Finland, Sweden, the Netherlands) is characterised by high levels of social protection expenditures (social and health care services which are mainly performed by the public sector and financed through taxation), high taxes, large public sector employment (more than 30% of total employment), basic social benefits for all citizens (universalism), an active labour market policy, compressed wage structures and weak employment protection legislation. The main problems with this model have to do with increasing difficulties in financing the costly welfare state under conditions of high capital mobility and the high need to expand private sector employment to compensate for the stagnation or decline in job opportunities in the public sector.
- The Anglo-Saxon cluster (United Kingdom, Ireland) is mainly characterised by social benefits for those in greatest need, increasing wage dispersion and a relatively large sector of low-pay employment. In contrast to the Nordic cluster, welfare state financing and private sector employment do not appear to be problematic. However, there is a lack of support (investment) for a highly competitive, highly skilled, export-oriented and well-trained labour force, thus resulting in significant levels of poverty and social exclusion.
- The Continental cluster (Germany, Austria, France, Belgium, Luxembourg) is characterised by unemployment and pension schemes primarily for those who have been in the labour market, a high degree of employment protection legislation and influential labour unions. In contrast to the Anglo-Saxon cluster, poverty and employment underqualification are not considered to be acute problems. However, there are clear signs of low employment levels in both the public and private sectors, high levels of benefit dependency and high long-term unemployment.
- The Mediterranean cluster (Greece, Italy, Spain, Portugal) is characterised by strict employment protection legislation, strongly compressed wage structures, and the understanding that social needs and risks are mainly covered by the family. This welfare state model has created a widening gap between labour market insiders with extensive benefits

and under-protected labour market outsiders. Similarly to the Continental cluster, the Mediterranean cluster is mainly financed through wage taxation, and private sector employment is priced out of the labour market. In terms of poverty, the Mediterranean model is similar to the Anglo-Saxon one.

Several studies (Ferrera, Hemerijck and Rhodes, 2000; Boeri, 2002; Sapir, 2005) have empirically compared the relative performance of the four models with respect to their meeting the most relevant objectives of social policy, and concluded that the Continental and Mediterranean ones are not sustainable due to their lack of economic efficiency. Due to the fact that the combined gross domestic product (GDP) of these countries accounts for almost two-thirds of EU GDP, it is obvious that this inefficiency is a problem for the EU as a whole. However, moving on to the Nordic or Anglo-Saxon models does not provide a "cure" because of the political costs associated with their governments as a result of citizens' strong opposition to the social imbalance they associate with a pure market economy.

The equation of the Social Market Economy is a balance among the answers to the following three questions: (a) how much "social" is needed? (b) how much "market" is allowed? and (c) how much government regulation is required to make the system successful? There is a constant debate concerning the need to adjust the equation of the Social Market Economy model. All along the way there have been those who argued for a greater or lesser governmental role, for more market or more regulation and for more social dimensions to be inserted into that equation, which leads to increasingly expensive development.

In today's globalised environment, it is crucial to identify the balance that encourages and requires the entrepreneurial spirit of the market while tempering its inherent propensity to run wild and concentrate more power at the top. Countervailing forces, in the form of a strong trade union movement, a diverse and healthy civil society and vigilant political parties, need to rein in the potential abuses and exploitation of capitalist practices by ensuring a just redistribution of the benefits of the market with the appropriate social programmes (Rifkin, 2004; Rifkin, 2005).

The Case of Greece

Greece has displayed a strong commitment to a European Social Market Economy model, and almost all political parties have supported the idea of a politically unified federal and social Europe.

All Greek governments have been in favour of further EU intervention on issues of employment, social inclusion, health and pension, the delegation of decision-making authority to EU bodies, and even further expansion of the welfare state through the establishment of new institutions and policies and the rise of social expenditure as a percentage of GDP.

This positive attitude towards European economic and political integration was fully justifiable because of (a) Greece's low economic development compared to that of most EU member countries; (b) the slow and insufficient development of social structures and the welfare state; (c) the wider consensus among the larger political parties on a European integration predominantly based on political criteria; and (d) the existence of a state-centred civil society and an inefficient public administration which have constrained the economic and social modernisation process, made problem solving difficult and negatively affected policy effectiveness overall (Sakellaropoulos, 2007).

With the Lisbon Strategy initially launched in 2000 and revamped in 2005, the European Union aimed at becoming the leading world economy in terms of both competitiveness and social cohesion. However, the renewed Lisbon Strategy actually weakened the goal of social cohesion relative to that of growth and employment by considering competitiveness and employment as preconditions for social cohesion.

Greece has always supported a political-institutional federal model of European integration, namely a strong, federal and politically united Europe, guaranteeing a strong economy and social protection system, regional cohesion, a democratic constitution, a common foreign and defence policy and a common fiscal policy. By entering the Eurozone, Greece introduced the euro and handed over its monetary policy to the European Central Bank. In a formerly unstable economy, without the danger of depreciation, the risk premium on Greek interest rates shrank to less than half a percent above that of Germany. This brought about convergence of Greek interest rates. Thus, investment and debt could be financed at lower rates. Member countries are fiscally sovereign regarding how to spend their taxes, but there is the Stability and Growth Pact in place, which has been set up to keep debt levels under control and to keep the currency area stable as a whole.

Greek governments used cheap refinancing conditions to expand heavily. At the same time, public sector employees realised wage growth that was far beyond productivity growth. Relatively cheap credit contributed to a booming economy and current account deficits against other European countries. Thus, the welfare gap between Greece and countries such as Germany appeared to close.

The Eurozone was greeted with serious doubts, mainly for economic reasons, at its inception. The current crisis and, more specifically, the efforts of the Eurozone to bail out Greece, reveals two more previously hidden arguments: politics and national culture. Greece sacrificed a bit of its sovereignty when joining the European Union but it appeared to be amply compensated by the substantial benefits it enjoyed. A bit more sovereignty was sacrificed when Greece joined the Eurozone, although this was not very noticeable, since it appears the government paid little attention to Eurozone strictures (Leontiades, 2010).

With a 15.4% budget deficit in 2009, Greece was clearly breaking the rules of the Stability and Growth Pact and worsening its refinancing attempts. Beginning in early 2009, interest rates on government bonds diverged strongly from other Eurozone countries, thus increasing the risk of bankruptcy. Nevertheless, Greece passed the initial inspections by the supervising body (the European Central Bank, European Union and International Monetary Fund) of its restructuring plan. In truth, there was no choice, as funds had to be made available to enable Greece to continue to function.

The changes proposed to Greece by the supervising body are definitely aimed not only at improving the functionality of the economy but also at bringing it closer to the German Social Market Economy model. The Eurozone was modelled on the discipline and institutions that made the German economy so successful. Adopting this model is less difficult for countries such as Belgium, the Netherlands and Luxembourg, which have similar institutional structures and cultural similarities. However, Greece is quite distant from the German model, not only in terms of culture but also in terms of political institutions.

The Case of Germany and the Rest of Europe

Germany, as well as most of its counterparts throughout Europe, is not particularly fond of market capitalism. There is a widespread distrust of the markets and a longing for a more just economic order which brings a better balance between economic growth and quality of life (Janes, 2010).

Today Germany, like its European partners, is coping with the need to slow down the spending side of the Social Market Economy equation. Controlling exploding levels of debt was a problem even before 2008, with even more debt building up as European governments continuously injected funds into their economies to stave off depression. The challenge of limiting social benefits constitutes a major obstacle in any government, and especially in Germany.

Most European citizens would prefer today a new economic framework that stimulates the development of a civil society which is not focused exclusively on economic growth and keeps a cautious eye on capitalism. However, the medium-term prospects of the Eurozone are quite bad and recovery will most probably be U-shaped, for many reasons (Roubini and Mihm, 2010). The projected growth rate of the Eurozone countries is low compared to that of the United States, Japan and the BRIC countries (Brazil, Russia, India and China). Furthermore, the Eurozone countries will face difficulties using fiscal policy to counter the effects of the crisis. Even before 2008, some of these countries ran large fiscal deficits and had high public debt relative to their GDP. These countries face several challenges over both the short term and long term, such as ageing populations and poor productivity growth. The PIGS countries (Portugal, Ireland, Greece and Spain) have experienced high debt and declining competitiveness. The adoption of the euro enabled them to borrow more and consume more than they would have otherwise, driving up wages and making exports less competitive.

The resulting mix of large current account deficits and budget deficits left the PIGS countries heavily indebted to banks elsewhere in Europe. The exposure of Germany and France to these four countries is approximately €730 billion, indicating that a default or a restructuring scenario would result in huge losses for the banks. In addition, the appreciation of the euro in 2008 increased the loss of competitiveness, leaving the PIGS countries even more vulnerable to default and threatening to burden the wealthier and/or healthier countries of the European Union.

In an effort to resolve the problems, the European Union announced a "rescue plan" which includes a €110 billion package for Greece (including the IMF's minority contribution), a €750 billion "safety net" for all Eurozone members, guaranteed ECB funding to vulnerable European banks and the ECB purchase of €60 billion of bonds issued by some troubled member countries.

Over the last decades, major economies around the world have been transferring debt from companies to consumers and ultimately onto the public sector. Reality has proven that the problem of extreme debt cannot – and should not – be solved with even more debt. There is a growing chorus of intellectuals within Europe who are echoing Einstein's statement: "You cannot fix a problem with the kind of thinking that created it."

The Day After

It would be unrealistic to assume that all the economic, financial and social challenges resulting from today's crisis will have a minor impact on people, their expectations, actions and fears. The damage to the quality of social capital may be particularly important. Tolerance of inequality, which has never been high in Europe, may be reduced further. Citizens may become more sensitive to social and economic divisions. Solidarity may also be weakened because of the many budgetary trade-offs, leading to contradictory budget demands and thus making budget consolidation even more difficult. Trust in public and international institutions will depend on the perception of their effectiveness and fairness in getting people safely out of the crisis, and most importantly, in putting economies and social services on a sustainable track over the longer term.

It is evident that the Social Market Economy model needs to be redefined. The financial and economic crisis which began in 2008 in the United States and quickly turned into a global depression is now rapidly becoming a social crisis, threatening to lead to a severe political crisis in some cases. It is apparent that all governments will, for some years to come, have to use extraordinary instruments of intervention to stimulate consumption, save the banking system as well as other industries and prevent social emergencies.

Nevertheless, the framework of principles according to which wealth is created, goods are distributed and human talent is put to use will not be changed. The market economy has a future provided it remains socially responsible, operating not as a goal in itself but as an instrument in the service of humankind (Martens, 2009).

To strengthen the Social Market Economy the EU should rapidly move forward by (a) including a legislative framework for bank crisis management, accompanied by proposals to reinforce protection for consumers of financial services as well as the regulation of credit rating agencies; (b) restoring job growth by incorporating new fiscal enforcement mechanisms, proposals to enhance the competitiveness of European small and medium-sized enterprises and rationalisation of the company taxation framework; and (c) improving citizens' everyday lives in the areas of justice, rights and freedoms by strengthening consumer rights, providing a Common Framework of Reference for Contract Law, improving the EU's Civil Protection Legislation and reorganising the EU's Anti-fraud Agency.

The implementation of the right mix of the Social Market Economy throughout Europe, though, is a difficult task because of cultural, political and economic restraints. The Social Market Economy model has been held up by some German politicians, academics and bureaucrats not only as the formula for Germany's success but also as a "cure for all ills". Yet this model assumes a process of consensus building which is very much part of Germany's political culture (Scharpf, 1999).

The Greek crisis depicts a much larger problem facing Europe, since governments appear to lack the political will to fight spiralling government debt. The willingness of governments to impose and citizens to bear the decline in living standards necessary to avoid a debt restructuring remains uncertain.

Governments must "invest" the necessary funds in a crisis to support the system and social balance, in line with Keynesian economics, but it is apparent that when the crisis is over they should revert to free market approaches, reflecting more of the Austrian approach to economics.

References

Boeri, T. (2002). *Let social policy models compete and Europe will win.* Paper presented at a conference hosted by the Kennedy School of Government, Harvard University, 11–12 April.

Esping-Anderson, G. (1990). *The three worlds of welfare capitalism.* Princeton, NJ: Princeton University Press.

Esping-Anderson, G. (1996). *Welfare states in transition: National adaptations in global economies.* London: Sage.

Esping-Anderson, G. (2000). A welfare state for the 21st century: Ageing societies, knowledge-based economies and the sustainability of European welfare states. Report prepared for the Portugese Presidency of the European Union, Spring.

Eucken, W. (1982). A policy for establishing a system of free enterprise. In Ludwig-Erhard-Stiftung, *Standard texts on the Social Market Economy,* 115–31. Stuttgart, New York: Gustav Fisher.

Ferrera, M., Hemerijck, A., and Rhodes, M. (2000). The future of social Europe: Recasting work and welfare in the new Europe. Report prepared for the Portugese Presidency of the European Union, Spring.

Hasse, R. H., Schneider, H., and Weigelt, K. (2008). *Social Market Economy history, principles and implementation – from A to Z.* Paderborn: Ferdinand Schöningh.

Janes, J. (2010). The debate over the Social Market Economy. *AICGS Advisor*, 19 August.

John, K. D. (2007). The German Social Market Economy – (still) a model for the European Union? *Theoretical and Applied Economics*, *3*(508), 3–10.

Leontiades, J. (2010). Will Greece leave the Eurozone? *Financial Mirror*, 30 April.

Martens, W. (2009). The future of market economy. *European View, 8*, 1–2.

Rifkin, J. (2004). *The European dream: How Europe's vision of the future is quietly eclipsing the American dream.* Cambridge: Polity Press.

Rifkin, J. (2005). Capitalism's future on trial. *The Guardian*, 22 June.

Roubini, N., and Mihm, S. (2010). *Crisis economics: A crash course in the future of finance.* New York: Penguin Press.

Sakellaropoulos, T. (2007). The political responses of eleven member states. In J. Kvist and J. Saari (Eds.), *The Europeanisation of social protection,* 211–27. Bristol: Policy Press.

Sapir, A. (2005). *Globalisation and the reform of European social model.* Bruegel policy brief 242.

Scharpf, F. W. (1999). *Governing in Europe: Effective and democratic?* Oxford: Oxford University Press.

Wrobel, R. M. (2010). Social Market Economy as alternative approach of capitalism after the financial and economic crisis. Paper presented at the 11th Bi-Annual Conference of European Association for Comparative Economic Studies (EACES), Estonia.

Political Leadership

Political Leadership in Greece in Times of Crisis*

Sir Michael Llewellyn Smith

There are many different types of leadership: charismatic, consensual, dictatorial and so on. And there has been much study, by political scientists, sociologists and others, of the theory of leadership in the varied spheres of politics, business and war. Most of what I write will be based on personal experience and observation. But I should like to start with a little theory as a backdrop against which to consider the qualities of leadership.

For the theory, I go back a long way, bypassing more recent academic studies and jargon. Here are some comments on political leadership by the great sociologist Max Weber (1864–1920), from his work "Politics as a vocation".[1] Weber famously wrote that a state is a "human community that (successfully) claims the monopoly of the legitimate use of physical force within a given territory." There is a clue there to one of the necessary qualities of political leadership: the ability and willingness to deploy force appropriately and responsibly where necessary.

On leadership, Weber wrote that "three pre-eminent qualities are decisive for the politician: passion, a feeling of responsibility, and a sense of proportion." He added that "the serving of a cause must not be absent … Exactly what the cause is … is a matter of faith. The politician may serve national, humanitarian, social, ethical, cultural, worldly, or religious ends … However, some kind of faith must always exist". Weber summed up with the thought that "politics is a strong and slow boring of hard boards. It requires both passion and perspective."

---

* This text is derived from a lecture Sir Michael Llewellyn Smith delivered at an event organised by the Konstantinos Karamanlis Institute for Democracy in Athens, on 30 June 2010.

[1] Quotations from Weber are taken from Gerth and Wright Mills (1948).

It will be evident that Weber was constructing an ideal, and that responsibility and passion were at the centre of his vision. It may be helpful also to look at the obverse of this ideal, what he calls the "deadly sins" of politics. These are "lack of objectivity" and "irresponsibility". He refers to the politician who is constantly in danger of being concerned merely with the impression he or she makes, whose lack of objectivity tempts him to strive for the glamorous semblance rather than the reality of power and whose irresponsibility suggests that he enjoys power merely for power's sake, without a substantive purpose. We can all think of such figures.

Looking at Greek politics over the last 150 years, I came up with three statesmen who seem to accord most closely with Weber's ideal, though of course they do not fit it in every respect: Konstantinos Karamanlis, Eleftherios Venizelos and Harilaos Trikoupis.

With Weber's comments and these three leaders in mind, I have constructed a diagram (table 1) showing the qualities of political leadership.[2]

Table 1. Qualities of political leadership

Qualities of political leadership			
Intellectual	Practical	Moral	And also …
Intelligence	Abilities (to manage, chair, delegate, persuade, communicate, get results)	Honesty (integrity)	Luck
Knowledge Judgement (proportion, objectivity, common sense)		Responsibility Vision	Good timing Magnetism
		Self-belief ("faith", "passion") Willpower Detachment Empathy	Stamina

[2] These qualities were established with the help of the audience and listed on a white board.

Some of these qualities are intellectual, some are practical, a few are physical and the most important are moral. The dividing lines between them are not clear-cut. "Judgement", for instance, crosses the boundaries. It is at the same time intellectual and practical and moral. Though political science can help in defining different types of leadership, assessing particular leaders is not a scientific business. It calls for moral judgements.

Some leadership qualities can be judged at the time. Others one can judge only in retrospect. Anyone can judge whether a Blair or a Brown, or a Karamanlis or a Papandreou, is the better communicator. But moral qualities, vision and judgement emerge through time. Posterity is the final judge, and even posterity can change its mind!

Also, the same leader may show different qualities at different times; leaders can become stranded in the past and unable to adapt to change, and thus lose their sense of judgement. Therefore, time and circumstance as well as luck are essential elements in successful politics.

I take as examples of successful leadership the years 1910 and 1974. Venizelos's achievement in 1910 and the succeeding three years, and Karamanlis's achievement in 1974 and the *metapolitefsi* are the two outstanding political achievements of the Greek twentieth century. In both cases leadership came in response to a real crisis. In the case of Venizelos, the immediate crisis was the blockage of normal politics caused by the Military League's uprising, followed by their realisation that they had messed things up. But behind that was a broader and longer-running crisis of national morale dating from the 1897 war. In fact much was achieved in the first decade of the twentieth century, in the economy and in the beginnings of military reform. But it did not seem like that to the junior officers of the Military League, who were fixated on the failures of nationalist and military policies in Macedonia and Crete.

In the case of Karamanlis in 1974, quite evidently the country was in a state of crisis after seven years of rule by the military junta and its sudden fall. So what did the two men bring to bear in order to steer the country back into "normal" democratic politics and transform the country?

In 1910–11 Venizelos showed political mastery of the first order in brokering the deal, involving the King, the Military League and the political parties, which enabled the country to resume business first under a provisional government and then, after elections, under Venizelos himself. He used constitutional change (the promise of a revisionary assembly) as a political mechanism for relaunching the country's politics. And then, cleverly, he brought about an important and lasting constitutional revision, including reforms such as permanence of tenure in the public service.

Venizelos was himself a "new man" in Greek politics, and he opened the way for a great injection of new talent into Greek political life. This is always difficult to do except in times of political rupture such as 1910–11 and 1974. I suspect many people today feel that there is a need for new blood and that the post-1974 political world has become blocked. But in the absence of such political rupture it is difficult to see how this can be achieved. Venizelos had a vision of where he wanted the country to go and faith in his own mission. He saw, and communicated, the need for the transformation of Greece, the modernisation of its institutions, the rebuilding of its economy and the expansion of its borders. He laid the groundwork for a modernised, liberal Greece within the European family of nations.

So he had many of the qualities we have identified: vision, faith, honesty, responsibility (in Weber's sense), willpower, intelligence, judgement, skills of communication and persuasion and management. He was not ruled by impressions or the glamour of politics. He also made mistakes – in one instance a mistake which had awful consequences. And the bad fairy gave him one attribute to balance all the good qualities: a divisiveness which was related to his conviction of his own rightness. This made him a difficult partner and even more difficult opponent.

Karamanlis in 1974 also had faith in his mission and his ability to reshape the country. He had a vision of where he wanted to take the country, a vision that matured over his period in exile in Paris. It included a historic political settlement with the Left, a historic constitutional settlement of the form of regime and a European destiny.

These are two examples of successful, transformative action, which would not have taken place without the personal stamp of the two individuals concerned. They fit well the concept of charismatic leadership described by Max Weber and elaborated for the Greek case by George Mavrogordatos.

I have been talking about particular periods of political creativity, when the man and the circumstances fitted each other. Of course a whole political career cannot sustain such achievement and success – as Venizelos's final years show. The British politician Enoch Powell said that "all political lives, unless they are cut off in midstream at a happy juncture, end in failure, because that is the nature of politics and human affairs." That is a superb aphorism – whether it is true or not is another matter – which illustrates a profound truth about politics, that politics is an ever-continuing process which can never end in success. Even the apparently successful career is subject to retrospective scrutiny, as for example in the critical assessments one reads of Karamanlis's post–Civil War building programme which transformed the face of Athens.

There were common features in the careers and the political thinking of Venizelos and Konstantinos Karamanlis. Both were outsiders: born outside the Greek kingdom. Venizelos learned his politics outside, too. Both believed at critical stages of their careers that the executive power in Greece should be strengthened. Both, like Trikoupis, spent periods outside Greece, in France, in a kind of exile. I suspect that all three of these things had a bearing on their leadership qualities and style as well on as their relationship with the Greek people.

Now I turn to the present day and offer some thoughts on how present circumstances bear on the question of political leadership. The political world here certainly believes that the country is in a state of deep crisis. Looking at the editorial pages of the issue of *Kathimerini* of 6 June 2010 I found the following: "Political system: Is there still any hope?" (Nikos Alivizatos); "The bankruptcy of politics" (Alexis Papahelas); and an article by Miranda Xapha which started with the sentence, "Sometimes I ask myself if Greece wants to be saved." I read also an article by Kostas Simitis, entitled "The price of a policy of clientelism", in which his opening words are "Greece presents the image of a country in deep crisis" and go on to refer to insecurity, pessimism and fear. I will return to this later.

So there is a deep sense of crisis. What sort of crisis? Clearly there is a crisis of public finances, in which politicians are struggling to avoid default and to keep Greece as part of the Eurozone. Finding the right response to this is a major challenge for both government and official opposition.

There are precedents in Greek history for such crises of public finances – for example Trikoupis in 1883 and Venizelos in 1932. But they are not much help to the authorities today because present circumstances are so different. The main difference is that Greece is locked into a single currency and is therefore at present deprived of tools available to prime ministers in the past (e.g., depreciation of the currency).

But behind this immediate crisis of debt, bond markets, austerity measures and so on lie questions about the nature of the political culture, the political system and the interaction of politics and society.

Note the question posed by Alivizatos, "Political system: Is there still any hope?" Does that mean that the political system itself is bankrupt? Bankruptcy as applied to politics is a metaphor meaning dysfunctional, devoid of ideas, incapable of dealing with crisis, compromised by involvement with scandals (Siemens, Vatopedi) and so on. There are some similarities between now and 1910, for example the sense of a malaise and pessimism in the country. But the political and economic circumstances are different.

The specific features of the present Greek "crisis" include

- a sense that the political world is inward looking and more concerned with party competition and electoral success than with the deep-seated problems of the country;
- a sense that necessary reforms – for example in education – have been blocked, that the system is incapable, whether for reasons of "political cost" or bureaucratic inertia, of driving necessary change; and
- a sense that the Greek political system has become stuck in the web of clientelist politics.

The article of Simitis to which I referred is instructive here. He talks of the "political compromise" on which the Greek political system rests. This compromise is based on the system of clientelism, on state handouts to interest groups and on avoidance of "political cost". He sees this system as intensifying the competition between political parties, destructive of honesty in politics and destructive of consistency and continuity in policy-making. Some of this, though not all, is simply part of the nature of democratic politics. And of course it happens in other countries too, including my own.

It is when Simitis comes to the solution that the problems appear. He writes that "we", namely all Greeks, should revalue time-honoured ways of thinking. The political system should no longer be limited to seeking ways of getting out of the economic crisis, but should abandon the clientelist policies of the past. Society should take responsibility for itself and stop pressing for preferential treatment at the cost of weaker members or groups. And "we should all agree to build a new collective interest for the country and our society."

All very good. But it is not clear where this new compromise is going to come from. It will not arise spontaneously within different social and interest groups. It calls, in a word, for political leadership of the highest order.

Simitis is calling for a change of mentality (*nootropia*). It is not clear how this is to be brought about. And when it comes to structural reforms – reform of the pension system, for example – government will come up against another systemic problem of the Greek political system, which is resistance to reform.

I read recently an article by a Greek political scientist working in England, Dimitris Sotiropoulos (2009), about obstacles to reform in Greece. He cited five different forms of failure of reform and nine causes. These include bureaucratic obstruction, weak state capacities, resistance by interest groups and elites and relations between the state and private interest groups which favour privileged groups at the expense of the people at large. Putting this in Venizelos's – or perhaps Simitis's – terms, the political world is polarised and the conception of the general interest is not strong enough to override claims on government of particular interest groups. So the way the political society operates and the legacy of history in the shape of client–patron relations are at the centre of the difficulty of achieving change.

What is the lesson of all this? Surely it is that deep and lasting changes will not come about simply through reform measures, however necessary, designed to take Greece out of the present crisis. First of all, those measures which get onto the statute book will not all work. Second, some of the underlying causes of blockage lie in the political culture itself: the nature of the Greek public service, the short span of the political cycle, the lack of ability and resources to implement decisions. Implementation is always as much a problem as legislative initiative and frequently more of a problem.

So it is all very difficult! But we knew that. We are brought back to Max Weber – "politics is a strong and slow boring of hard boards" – calling for high qualities of political leadership. Does it have to be normal politics then, business as usual with all the features condemned by Simitis and others? Not necessarily. There have been a number of suggestions for improvements that could be made.

One is constitutional change. When there is a crisis people tend to call for amendments to the constitution. But changes to the constitution do not themselves necessarily solve political problems. They help only if they are part of a leader's vision of what needs to be done, as they were in 1911 and 1975.

Another is to make better use of the political and intellectual forces of the country. I don't mean so much coalition or ecumenical government, though there is a place for that in certain circumstances, in Greece as in Britain, as for government of its own accord to make better use of expertise from outside its own party circle.

What I have argued is that behind the present crisis and linked to it are deep-seated, long-established, structural problems of institutions, of society and of political culture. These are for the whole of civil and political society, politicians, media, intellectuals, teachers, doctors, workers and so on, to grapple with. And the lead needs to come from the top.

That brings us back to the qualities of political leadership. As always, specific political skills will be required – skilful party and human management, a sense of timing, the ability to win and keep power and so on. But the problems of these times also call for moral and intellectual qualities: knowledge not just of finance and economics but also of history, society and culture; good judgement; and a vision of the desired ends, based on an appreciation of what is wrong. It is not only Greece that requires such leadership.

References

Alivizatos, N. (2010). Political system: Is there still any hope? (in Greek). *Kathimerini*, 6 June. Available at http://news.kathimerini.gr/4dcgi/_w_articles_columns_2_06/06/2010_403553, accessed 22 November 2010.

Gerth, H. H., and Wright Mills, C. (1948). *From Max Weber: Essays in sociology*. London: Routledge.

Papahelas, A. (2010). The bankruptcy of politics (in Greek). *Kathimerini*, 6 June. Available at http://news.kathimerini.gr/4dcgi/_w_articles_columns_2_06/06/2010_403559, accessed 22 November 2010.

Simitis, K. (2010). The price of a policy of clientelism (in Greek). *Kathimerini*, 2 May. Available at http://news.kathimerini.gr/4dcgi/_w_articles_politics_2_02/05/2010_399578, accessed 22 November 2010

Sotiropoulos, D. A. (2009). The paradox of on-reform in a reform-type environment: Lessons from post-authoritarian Greece. Paper delivered at Yale University, now awaiting publication.

Xapha, M. (2010). Abusive privileges and populism (in Greek). *Kathimerini*, 6 June. Available at http://news.kathimerini.gr/4dcgi/_w_articles_columns_2_06/06/2010_403546, accessed 22 November 2010

In the Name of "Europe": Analysing Prime Ministerial Discourse from EU Membership to the Greek Financial Crisis

Christos Dimas

Introduction

This chapter intends to outline the evolution of the employment of "Europe" in Greek prime ministerial communicative discourse over the years. It aims to illustrate how the current Prime Minister, George Papandreou, has adopted a double discourse in matters involving Europe. The first is the traditional domestic discourse in which governments anticipate garnering increased public justification for implementing their policies by legitimising their choices in the name of Europe, especially when proposed policy reforms are contentious and the opposition grows. The second, which seems to be a more recent phenomenon, has to do with the external discourse and the unconventional effort that the current Prime Minister has made to convince Europe and the international financial markets that Greece is changing and that it will successfully apply the austerity measures that Europe has enforced.

Linking the Institutional Context to Justification in the Name of Europe

Since 1977, when the Greek application to the European Economic Community (EEC) was discussed in the national Parliament, major decisions in Greek politics have been largely justified by national actors in the name of Europe. In 1977, Greek Prime Minister Konstantinos Karamanlis interrupted the speech of the leader of the Opposition, Andreas Papandreou, who had accused the government of sacrificing the country on the altar of "we belong to the West". The Prime Minister argued that Greece belongs to the Western European world both by tradition and interest. Three decades later, Prime Minister George Papandreou addressed the nation and outlined the reasons why the country had to resort to activating the EU-IMF economic support mechanism which would eventually result in a series of austerity reforms. Once more, a Greek Prime Minister systematically referred to Europe and stated (BBC, 2010) that "our European partners will decisively contribute to provide Greece the safe harbour that will allow us to rebuild our ship."

The main argument of this chapter is as follows: the national institutional context has not incorporated interest groups within a functional corporatist system and has been largely characterised by ongoing political and social polarisation. As a result, national governments have acted unilaterally and have attempted in their communicative discourse to justify their policy choices by appealing directly to the general public. According to Schmidt (2008, pp. 310–11) communicative discourse "consists of the individuals and groups involved in the presentation, deliberation, and legitimisation of political ideas to the general public." Governments and prime ministers personally seek to follow a particular discourse that will simultaneously justify their choices and, more importantly, will facilitate their efforts to build impetus and garner adequate public support to implement their policy choices. The objective is to win over as much of the general public as possible in order to legitimate the application of their policy choices. Hence, they decide to associate the application of a policy with an external stimulus, usually the country's European requirements and/or objectives, thus hoping that the general public will show more understanding for their application. As Tsoukalis (2000, p. 42) comments, "EU policies and rules can sometimes serve as convenient scapegoat for unpopular policies at home. Greek governments have frequently made use of the European scapegoat whenever domestic support was short in supply. They have tried to capitalise on the generally high levels of public support for the EU at home."

Discourse in the 1970s and 1980s: Nea Demokratia Versus PASOK, or Euro-Enthusiasts versus Euro-Sceptics

Ever since the reestablishment of democracy in 1974 and up until the mid-1990s, there were two vital ideological differences that characterised the two main political parties, Nea Demokratia (ND) and the Panhellenic Socialist Movement (PASOK). The first concerned the international status of the country and whether Greece should join the EU and NATO, and the second was based on the issue of liberalism versus socialism. Both debates were intense and polarised the political and social climate to a great extent until the mid-1990s.

Since the founding of the two political parties in 1974, the debate concerning prospective EEC membership has been one of the most divisive issues between their leaders. ND's founder Konstantinos Karamanlis was a genuine supporter of an economically and politically united Europe and immediately made it clear that ND had an unambiguous European orientation. This was verified by the party's founding manifesto (ND, 1974), which declared that

> ND believes that Greece not only has the right, but can actually safeguard the pride and the happiness of the people within Europe, where it belongs, if it makes sure to mobilise all the abilities and virtues of its people. Independently of its size, Greece's cultural heritage, the Hellenic aura and the spirit of the Greek people can assist Europe politically, ethically and culturally in order to complete the European union.

In contrast, PASOK's ideological position was to directly oppose EEC and NATO membership. This was clearly outlined in its founding manifesto (PASOK, 1974), which proclaimed that "Greece should withdraw from NATO … Greece should detach itself from any military, political and economic alliances that undermine our national independence and the right of Greek citizens to decide for themselves concerning social, economic and cultural aspects of life."

Due to the two successive electoral victories of Karamanlis's ND in 1974 and 1977, the debate on Europe tilted in the direction set by the ND governments. In fact, after the 1974 elections, Prime Minister Karamanlis pointed out that EEC membership was the primary goal of his government, mainly for reasons of safeguarding democratic consolidation. He believed that membership would sustain Greek interests within a powerful, democratic economic community. More importantly, his diplomatic manoeuvres achieved an early conclusion of the accession agreement, which was signed in May 1979. PASOK criticised the government and described EEC membership as the surrendering of Greek interests to the objectives set by unwanted foreign powers.

Two very important political developments, which eventually proved to conflict with each other, sparked change in the Greek political arena in 1981. First, Greece became the tenth member of the EEC. In fact, Greece's future partners did not accept its application with much warmth as the Commission had advised the Council not to accept the application on economic grounds. Second, Andreas Papandreou's charismatic personality, his well-known family name and socialist jargon made him the only alternative to a ND government, as he led his party to an electoral landslide in 1981.

The combination of Greece having entered the EEC and Papandreou's lack of eagerness to adjust to EEC policies led to a troublesome relationship between the Community and Greece during the 1980s. Papandreou, who had opposed membership, found himself within the European Council. He had repeatedly argued that Greece should disengage itself from foreign alliances that neglected Greek national interests in favour of those of foreign powers. However, in time Papandreou's radicalism receded and several contentious issues, especially regarding the international status of the country and its withdrawal from the Community, were gradually abandoned.

In any case, due to the government's economic policy, by 1989 an oversized public sector had formed which was characterised by an ineffective and inefficient modus operandi. Thus it is no surprise that Greece under Papandreou was regularly portrayed as the "awkward member" (Papadopoulos, 2004) or the "political and economic black sheep of the EU" (*International Herald Tribune*, 1999). Dinan (1994, p. 83) argues that "if the EC could have foreseen the problems that Greek membership would pose in the 1980s and early 1990s during the rule of Andreas Papandreou's anti-EC governments, the accession negotiations might not have concluded so swiftly, if at all."

The constant mismanagement of the Greek economy was an issue which the Community could not continue to ignore. As Featherstone (2003, p. 933) points out, "in March 1990, the then Commission President, Jacques Delors, wrote to Xenophon Zolotas, the technocratic head of the all-party government, warning that the deteriorating economic situation in Greece was 'a serious concern for all of us'." Indeed, the dire Greek situation threatened the ability of the EC to achieve its major common objectives: the single market, Economic and Monetary Union (EMU) and the unification process as a whole.

The Europeanised Discourse of the 1990s and 2000s

After it second consecutive electoral defeat at the hands of PASOK in 1985, the leader of ND, Konstantinos Mitsotakis, underlined that his party would win the next elections only if it convinced the public that it had been transformed into a modern European political party with a manifestly neoliberal economic agenda. As a result, in June 1987 ND presented its new economic programme which was not only in deliberate harmony with Community objectives but was actually built around them.

The party established as its highest priority the country's effective preparation for the single market. ND stressed that it was wholeheartedly in favour of the single market and considered adjustment to Community requirements and legislation as the only way forward to modernise the public sector and improve the national economy. This was spelled out clearly in the party programme (ND, 1987, p. 14) which stressed that

> the Single European Act has opened the way for the completion of the European internal market. In a few years' time it will be impossible to maintain institutions that are anachronistic and non-productive. ... The vision of 1992, which Greece has also espoused, is a truth and therefore ND has incorporated in a harmonious and consistent manner all the necessary measures, institutional and others, so that the country may live up to the great challenge.

One of the most evident examples of the employment of Europe as a tool to justify domestic policy choices is the first wave of the Greek privatisation programme which took place during the ND government of Konstantinos Mitsotakis from 1990 to 1993. In its three years in office, the Mitsotakis government placed privatisation at the centre of the political agenda. The Prime Minister, in his communicative discourse, systematically linked the application of privatisation to the advantages of belonging to the Community through a series of manifestos, press releases, speeches and interviews. Characteristically, Mitsotakis (1992) declared that "for us the EU is our central national aim and our main pursuit. This is why we will accelerate the privatisation process, which I have to confess is a difficult matter, but recently our results are very positive."

The neoliberal, European-inspired strategy adopted by the party and its Europeanised discourse helped to serve a triple goal. First, it presented the Greek public with an alternative ideological foundation that would help modernise the unproductive public sector in contrast to the socialist model advocated by PASOK. Second, it provided ND with the necessary macroeconomic policies, tools and external discipline for the adoption of internal austerity reforms. And, finally, as ND directly linked neoliberalism with the single market, it was regarded as being a part of a pan-European initiative based on common principles and objectives and not solely on a domestic policy decision which sought to accomplish short-term political benefits. Nevertheless, despite the government's European discourse and neoliberal agenda, its fragile parliamentary majority did not allow it to last more than three years, and therefore it was not able to apply its programme.

Before the 1993 elections Andreas Papandreou expressed his party's intention to cooperate with the Community in economic policy and erase the anti-EU reputation he had gained as Prime Minister in the 1980s and had reinforced when he was in opposition. One month before PASOK's electoral triumph, on 3 September 1993, exactly 19 years after the publication of its founding manifesto, Papandreou presented a revised version. The party's propensity to be an EU sceptic was limited compared to the past. Thus the revised manifesto did not include any polemical statements against EU and NATO membership. In contrast, it acknowledged the importance of supranational institutions for a well-functioning economy and enhanced national security. Papandreou actually stated that his party had embraced Europeanisation (Papandreou, 1993) because "the future of our country in Europe, as a country that fully participates in the European evolution, will depend on the policy followed by the next government that will be formed after the elections." Papandreou was not able to complete his term, being forced to resign in January 1996 due to health deterioration.

From the beginning of his term as Prime Minister, Kostas Simitis led the Europeanisation campaign in a rigorous communicative discourse informing the public of the possible consequences if Greece did not fulfil EMU criteria on time. Simitis regularly cited prospective EMU participation as the main reason justifying the implementation of the restructuring of the public sector. He was very effective in delivering his message to the general public and in explaining that Greece's failure to participate in the EMU would be damaging to the well-being of the national economy. The Prime Minister (Simitis, 2005, p. 169) noted that

> PASOK's new government simplified and expressed the dilemma that the country was facing at the time: do we wish to be part of the powerful global economic centres and have the capability to influence a wide area of policies and have a solid currency thus arming the national economy against international crises? Or

do we believe that despite our shortcomings we will be capable, on our own, to control international developments to our advantage? EMU is the means to put an end to the times when Greece was a peripheral member of the Union. We should not allow our country to miss out on any future opportunities.

The government publicly and successfully supported EMU membership in its communicative discourse and in turn carried out pro-market reforms such as market liberalisation, deregulation and privatisation. Simitis argued that they were the keys to EMU participation, which was acknowledged as the great national objective. Characteristically, the Prime Minister (*Kathimerini*, 1996) stated that

by all means we must succeed in being part of the core EU Member States in the EMU. Only then will we be able to have an influential role in the decisions that will be affecting us. Therefore EMU membership is not only an economic issue, but mostly a political one. In fact the economic policy that we have decided to apply is part of an entire development plan which aims to restructure all of the public sector.

The ND government which was elected to office in 2004 regularly referred to Europe in its domestic discourse as a way of highlighting the legitimacy of its chosen policies. Prime Minister Kostas Karamanlis demonstrated this when the government, Marfin Investment Group and Deutsche Telekom signed the deal which established the latter as a shareholder in Hellenic Telecommunications Organization (OTE). Karamanlis sent a message to the Greek people by indicating that this strategic alliance would help OTE become more competitive in the European market. More specifically, he noted (ND, 2008) that

with regards to what is taking place in Europe, it is obvious that the opposition is either unaware of what is happening in Europe or it is pretending not to know. Therefore, I make it clear that most European countries have proceeded with the entire privatisation of their telecom providers, such as in Spain, UK, Ireland, the Netherlands and Portugal. Furthermore, two of our European partners (Sweden and Finland) have a common telecom company.

However, when it came to privatisation, even if the Karamanlis governments still referred to Europe as the reason legitimating it, the policy was not as divisive as it had been in previous years, since the state by that time had already sold many state-owned entities. In addition, ND held a strong ideological position in favour of privatisation. Therefore it did not need to justify its policies on other grounds.

2010: The Double Discourse of the Prime Minister

On 4 October 2009, George Papandreou's PASOK took office after winning the general elections. PASOK's campaign had been structured around its intention to give marginal increases – just above the inflation rate – to the public sector employees and its promise to renegotiate deals signed by the previous government such as those with Olympic Air, OTE, the port of Piraeus and the energy sector. Nevertheless, the newly elected Prime Minister quickly found himself facing the Greek financial crisis and naturally had to back down from most of his pre-election pledges. Papandreou enacted an external discourse which involved a series of speeches abroad and interviews given to the foreign press in which he rigorously accused the previous government and blamed it for the country's predicament. He argued (Papandreou, 2010b) that "there were certain endemic problems, but, in the last five or six years, unluckily, the previous government, rather than dealing with these endemic problems, exacerbated them." In addition he repeatedly emphasised that Greece was a very corrupt country. He stated (Elliot, 2010) that "the problem we have is home-made … we developed a lot of corruption at the highest levels and we did not take the structural measures to change our economy, to move our economy, to make it more competitive." Finally, he made frequent remarks claiming (Papandreou, 2010a) that "the country's main problem was not the budget deficit, but the credibility deficit." It has been argued that this external discourse was partly responsible for the magnification of the Greek problem, as Papandreou constantly drew a depressing picture of the situation in Greece and thus contributed to the problem rather than helping to deflate it.

The double discourse became more evident than ever when the Prime Minister addressed the general public in April 2010. He announced the decision to resort to the EU-IMF bailout mechanism. At that stage Papandreou's wording was very careful, as he was obliged to balance his discourse in order to keep content Greek citizens and at the same time send the right signals to Greece's EU partners and the financial markets. On the one hand he continued to blame the previous government and the greediness of the financial markets while on the other hand he painted a more optimistic picture for the future based on the assistance that Europe would provide. He underlined the importance and the critical role that Europe played and stressed that under the guidance of its European partners

Greece would find its way to fiscal stability. More specifically, Papandreou (2010c) stated that

> Greece had become a country lacking status and credibility and had lost its respect even from its friends and partners … we [this government] asserted and managed to seal a good deal with the EU in order to support our country … We and our partners in the EU hoped that the decision to initiate the mechanism would be enough to calm and confront the markets with their responsibilities so that we could continue to finance the country with lower interest rates … this did not happen and since the time we needed was not given to us by the markets it now will be given by the EU. Thus we have formally asked our EU partners to activate the support mechanism.

The Prime Minister was forced to change his discourse in order to convince both Greece's European partners and the international markets that Greece would implement successfully its measures and would be in the position to pay back its loan instalments in the future. This did not mean that the double discourse was abandoned, however, but only that its content differed. Characteristically, he pointed out that "we, in Greece will do everything that is needed to be done. Europe will see that we are actually changing" (Papandreou, 2010e). It is striking that on various occasions he found himself in the uncomfortable position of having to defend even the diligence and credibility of the Greek people. The German media in particular were very critical and led Papandreou (2010d) to state that

> in Europe there are over-reactions, prejudices and many stereotypes against the Greek people. Many Europeans know us only from their holidays and what they see is people that love life. We do, but this is only one side of the truth. The other is that we Greeks work hard. You, in Germany, should know that well as there are hundreds of thousands of Greeks who have worked as labourers in German factories. Often, the Greeks were those that did the jobs that nobody else would do. Many of those migrants later formed their own businesses or followed an academic career.

Conclusion

Indeed policymakers in Greece have appealed to Europe in their domestic discourse to justify their proposed reforms within the country. They have actually shown that when things get rough, they resort to and anticipate that Europe will – either practically or theoretically – be there to redeem them. When commenting on the effects of the global financial recession and the EU-IMF bailout mechanism for Greece, Professor Ferguson of Harvard University (Ferguson, 2010) stated that "in desperation, the Greeks turned to their fellow Europeans for assistance." However, this is not necessarily a recent development, since one of the main reasons that the Greek governments of Konstantinos Karamanlis strongly held that the country should apply for EEC membership was in order to safeguard democratic consolidation in the country, thus protecting it from internal and external threats. The assumption that Europe will save Greece from difficult situations is not a recent phenomenon.

In conclusion, there have been two forms of prime ministerial discourse evident in Greek politics since the emergence of the financial crisis. First, there is the traditional form of discourse which takes place when the government's proposed reforms are likely to touch what the public sector or other well-organised interest groups consider to be their "vested rights." Due to the polarised climate and the lack of a corporatist system, the latter unite and raise their voice against the reforms. As a result, consecutive governments have attempted to bypass this hurdle unilaterally, traditionally by appealing to the general public and by using Europe as their ultimate argument in order to increase social justification and overcome domestic blockages of interests regarding their proposed policies. Second, and more interesting, the most recent form of external discourse also includes a European element. It differs from the traditional one in the sense that the Prime Minister does not address national citizens but Greece's EU partners and the international financial markets. It signifies an attempt to convince them that Greece is on track, going in the right direction as it successfully implements the policies that have been prescribed by the EU itself and the IMF. At the same time, however, it illustrates that when the external discourse is misused it may create more problems than policymakers think it will solve. In relation to the first type of discourse, history has proven that the appeal to Europe may in some circumstances be a useful tool indeed, but it is not sufficient on its own to gather the necessary social legitimation to successfully apply the policies at home. As for the second type of discourse, it remains to be seen whether it will have any other noticeable and practical effects on the relations between the Greek government, its EU partners and the international financial markets.

References

BBC. (2010). Greece calls on EU-IMF rescue loans. 23 April. Available at http://news.bbc.co.uk/2/hi/business/8639440.stm, accessed 15 October 2010.

Dinan, D. (1994). *Ever closer union: An introduction to European integration*. Houndmills: Palgrave.

Elliot, L. (2010). No EU bailout for Greece as PM promises to "put house in order". *The Guardian*, 28 January. Available at http://www.guardian.co.uk/business/2010/jan/28/greece-papandreou-eurozone, accessed 15 October 2010.

Featherstone, K. (2003). Greece and EMU: Between external empowerment and domestic vulnerability. *Journal of Common Market Studies*, *41*(5), 923–40.

Ferguson, N. (2010). The end of the euro. *Newsweek*. 17 May.

International Herald Tribune. (1999). 13 January.

Kathimerini. (1996). 20 November.

Mitsotakis, K. (1992). Speech at the Greek–American Chamber of Commerce. Athens, 22 May. Available at www.ikm.gr, accessed 15 October 2010.

ND (Nea Demokratia). (1974). *ND founding manifesto*, 4 October. Available at www.nd.gr, accessed 15 October 2010.

ND (Nea Demokratia). (2008). Answer of Prime Minister Kostas Karamanlis to a written question submitted by the Leader of PASOK, George Papandreou. Prime Minister's Office, 16 May. Available at http://www.nd.gr/index.php?option=com_content&task=view&id=49681&Itemid=486, accessed 15 October 2010.

ND (Nea Demokratia). (1987). *Economic programme: A free-market, competitive and democratic economy, Secretariat of the party's programme* (in Greek), June 1987.

Papadopoulos, C. A. (2004). Greece and the EU at Helsinki: A historical new departure in Greek–Turkish relations? Euroborder Conference. Available at http://www.euborderconf.bham.ac.uk/case/GreeceTurkey/Gr-TPapadopoulos.pdf, accessed 15 October 2010.

Papandreou, A. (1993). Interview with Flash 96.1 FM, 27 September. Available at www.pasok.gr, accessed 15 October 2010.

Papandreou, G. (2010a). Interview with ABC and *Lateline Show* with journalist Philip Williams, 24 February. Available at http://www.primeminister.gr/2010/02/25/965, accessed 15 October 2010.

Papandreou, G. (2010b). Interview with PBS, 8 March. Available at http://www.pbs.org/newshour/bb/business/jan-june10/greece2_03-08.html, accessed 15 October 2010.

Papandreou, G. (2010c). Speech at the Government meeting in Kastelorizo, 23 April. Available at http://www.primeminister.gr/2010/04/23/1623, accessed 15 October 2010.

Papandreou, G. (2010d). Interview with newspaper *Handelsblatt* and journalists Gerd Hohler and Gabor Steingart. 17 May.

Papandreou, G. (2010e). Interview with newspaper *Die Zeit*, 30 September. Available at http://www.primeminister.gr/2010/09/30/3124, accessed 15 October 2010.

PASOK. (1974). *PASOK founding manifesto of 3 September 1974*. Available at www.pasok.gr, accessed 15 October 2010.

Schmidt, V. A. (2008). Discursive institutionalism: The explanatory power of ideas and discourse. *Annual Review of Political Science*, *11*, 303–26.

Simitis, K. (2005). *Policy for a creative Greece, 1996–2004* (in Greek). Athens: Polis Publications.

Tsoukalis, L. (2000). Greece in the EU: Domestic reform coalitions, external constraints and high politics. In A. Mitsos and E. Mossialos (Eds.), *Contemporary Greece and Europe,* 37–51. Aldershot and Burlington: Ashgate.

Climate Change and Environmental Protection

Climate Change: An Issue of International Concern

Christos Zerefos

Introduction

The earth's climate is changing because of natural and anthropogenic contributions to the environment's composition and structure. A balance has been maintained in our environment by interactions continuously operating since the dawn of life, between the biosphere, the atmosphere, the hydrosphere, the cryosphere and the geosphere. These five spheres are continuously interacting to maintain a quasi-ecological equilibrium which includes adaptation by humans and ecosystems to new states of environmental equilibrium. In the past few decades, which form the so-called anthropocene period, it has been recognised that humans are contributing to climate change through a number of activities, the most notable of which are anthropogenic global emissions of greenhouse gases.

Anthropogenic climate change is part of an overall global change which is the result of the natural environmental interactions mentioned above. This has significant impacts on biological and physico-chemical systems and on humans. The effects are distributed globally and are expected to have important consequences not only for humans but for the ecosystem as well. Among the most prominent of these consequences are changes in the hydrological cycle on both global and regional scales, changes in the intensity of extreme weather events and changes in rainfall characteristics and patterns among them. These changes are expected to affect the incidence and severity of droughts and floods, making problematic in some cases the availability of water and food production.

Global changes will result in challenges for health and many other aspects of human societies, for industry and the overall wealth and security of the planet. Notable are the foreseen effects on economies, agriculture and food and water security. The list of negative consequences from anthropogenic global change is indeed large: from sea-level rise, with severe effects on human settlements, to the observed effects on biological systems. There is evidence of changes and shifts in the range of plant and animal species to higher latitudes and altitudes and changes in the timing of many life-cycle events, such as flowering. This variability will affect ecosystems and biodiversity in general. Many of these impacts, especially when operating in synergy, are expected to cause additional threats to humans, to our resources and to ecosystems.

The interactions among the hydrosphere, the atmosphere, the biosphere, the cryosphere and the geosphere have been operating on earth for millions of years. Recent concerns about the effects of human activities that intervene with natural globally changing processes are based on observation, modelling and statistical analysis. Important international efforts aim to integrate these observations and coordinate an international network of modellers and experimenters in global-change research.

Discussion

The year 2010 has been dedicated to the biodiversity of our planet. It is the year that great decisions have to be taken globally. These decisions have to create the path towards our liberation from dependence on conventional fuels and other greenhouse gases. It is only in the past few decades that scientists have recognised beyond any doubt that emissions from fossil fuels and the production of other anthropogenic greenhouse gases have serious effects on the global environment of our planet. This is because greenhouse gases can alter the radiation balance in our environment leading to a general destabilisation of our climate system. This destabilisation will trigger more frequent extreme events in our interactive atmospheric hydrosphere system and impose regional phenomena with significant costs to the economy.

Recent studies have shown that extreme weather events have been particularly modified, presumably by the anthropogenic destabilisation of the climate in the Mediterranean. A new dataset of high-quality homogenised daily maximum and minimum summer air temperature series from 246 stations in the eastern Mediterranean region (including Albania, Bosnia-

Herzegovina, Bulgaria, Croatia, Cyprus, Greece, Israel, Romania, Serbia, Slovenia and Turkey) was developed and used to quantify changes in heat-wave frequency, length and intensity over the past 50 years. Daily temperature homogeneity analyses suggest that many instrumental measurements in the 1960s are warm-biased; correcting for these biases, regionally averaged heatwave trends are up to 8% higher. It was found that significant changes occurred across the western Balkans, south-western and western Turkey, and along the southern Black Sea coastline. Since the 1960s, the mean heat-wave intensity, heat-wave length and heat-wave frequency across the eastern Mediterranean region have increased by a factor of 7.6 ± 1.3, 7.5 ± 1.3 and 6.2 ± 1.1, respectively. These findings suggest that the heat-wave increase in this region is higher than previously reported (Kuglitsch et al., 2010). There is also evidence of significant changes in the probability distribution of extreme rainfall in Athens and elsewhere in the Mediterranean (Nastos and Zerefos, 2008, 2009).

The changes in daily precipitation in Greece over a 45-year period (1957–2001) have also been examined. The precipitation datasets concern daily totals recorded at 21 surface meteorological stations of the Hellenic National Meteorological Service, which are uniformly distributed over the Greek region. First and foremost, the application of factor analysis resulted in grouping the meteorological stations with similar variation in time. The main subgroups represent the northern, southern, western, eastern and central regions of Greece with common precipitation characteristics. For representative stations of the extracted subgroups we estimated the trends and the time variability for the number of days (%) exceeding 30 mm (equal to the 95 percentile of daily precipitation for eastern and western regions and equal to the 97.5 percentile for the rest of the country) and 50 mm (which is the threshold for extreme and rare events).

The European Union and the European Space Agency have initiated the Global Monitoring of the Environment and Security (GMES). This activity, named Kopernikus, aims to provide to EU countries an operational service for studying anthropogenic changes in the environment and their related impact on the security of citizens. However, the synergy between slowly and rapidly varying components of geophysical extremes, enhanced by the anthropogenic change to the environment, is an issue of urgency, having several missing links not previously considered in such an operational activity. This is particularly important when we investigate the synergistic effects of disasters caused by nature and accelerated by humans, and their effects on humans, society and ecosystems. For example, earthquakes can be linked with landslides and tsunamis in coastal areas. The synergistic effects of these natural disasters in an anthropogenic, globally changing environment are today mostly unknown and, worst of all, there is no existing European infrastructure dealing with interactions between different types of extreme events occurring in synergy under the laws of probability. The Mediterranean region is vulnerable to natural and anthropogenic disasters. Natural disasters have always occurred in this part of the world, but anthropogenic changes to our environment have worsened the effects on humans and on the ecosystems of natural disasters.

The Mediterranean is already under pressure from global stresses and is highly vulnerable to the impacts of climate change (Luterbacher et al., 2006). Floods and droughts can occur in the same area within months of each other. These events can lead to famine and widespread disruption of socio-economic well-being, particularly along the North African shore. Many factors contribute to the impact of anthropogenic climate variability in the Mediterranean, making it difficult to cope with these changes. The over-exploitation of land resources including forests, an increase in population, desertification and land degradation are additional threats in this area (UNDP, 2006). In parts of the Mediterranean and along the North African shore, dust and sand storms have negative impacts on agriculture, infrastructure and health. The Mediterranean is also expected to face increasing water scarcity and stress. Agricultural production relies mainly on rainfall for irrigation and will be severely compromised. As a result of climate change some agricultural land will be lost, with shorter growing seasons and lower yields. Rising temperatures are changing the geographical distribution of disease vectors, which are migrating to new areas and to higher altitudes and latitudes (WHO, 2004).

Climate change is an added stress to already threatened habitats, fragile ecosystems and species in the Mediterranean; land-use changes due to agricultural expansion and the subsequent destruction of habitat, pollution, high rates of land-use change, population growth and the intrusion of exotic species are likely to lead to habitat reduction and will trigger species migration. In addition, climate change is altering weather and climate patterns that previously have been relatively stable. Climate experts are particularly confident that climate change will bring increasingly frequent and severe heatwaves and extreme weather events, as well as a rise in sea levels. These changes have the potential to affect human health in several direct and indirect ways, some of them severe.

Heat exposure has a range of health effects, from mild heat rashes to deadly heat stroke. Heat exposure can also aggravate several chronic diseases, including cardiovascular and respiratory disease. The results can be severe and lead to increases in the number of illnesses and deaths. Heat also increases ground-level ozone concentrations, causing direct lung injury and increasing the severity of respiratory diseases such as asthma and chronic obstructive pulmonary disease. Higher temperatures and heatwaves increase demand for electricity and thus the use of fossil fuels, generating airborne particulates and indirectly contributing to increased respiratory disease.

Hot days and heatwaves present another hazard that can act in synergy with other events. This happened, for example, during the heatwave in Europe in August 2003. There were more than 40,000 deaths. This heatwave was created by a confluence of specific meteorological events and resulted to increased number of deaths in Paris, Torino and Barcelona in very high correlation with the air temperatures reported during the heatwave. The situation here was a common experience in a period of about two to three weeks; most of the victims were elderly people who were not prepared and not properly informed how to deal with the situation. Elderly people often take drugs which dehydrate them. So without any warnings or any information provided to them, they became dehydrated, were exposed to heatwave stress and, unintentionally, to greater danger of death. This is one of the most extreme or worst examples of danger induced by synergistic effects (heatwave, high ozone levels, high dehydration, limited knowledge, no warnings to society). A model prepared by the University of East Anglia has shown that the heatwave of 2003 will be considered in the future as a "cool" event when compared to the mean temperatures beyond the year 2060.

The history of heatwaves is as long as the history of our planet, but the severity of their effects was shown to be at its peak, in terms of health, during the 2003 event. We should note that in addition to the 2003 disaster Europe experienced a striking heatwave this year (as well as in previous years; for example, in 2007), which, fortunately, because of better warning and better education of society, did not result in the death toll seen in 2003.

Heatwaves can create a high risk of forest fires. The recent disaster in Russia and previous disasters in California and the Mediterranean are good examples of the risky environment that results from extremely dry atmospheric conditions. These extreme events occur, as we know, in the summer time, and particularly in the Mediterranean one can see that this area is also extremely vulnerable to very high levels of ozone, which is also an respiratory threat (Zerefos et al., 2002). So oxidants in the atmosphere, in addition to causing dehydration and heat stress, can result in deaths which could be prevented; they have resulted also in creating the set of rare conditions for the re-occurrence of forest fires, which also operate in synergy, threatening humans, ecosystems and biodiversity.

Over longer periods of time, increased temperatures have additional effects on health. Droughts can result in shortages of clean water and may concentrate contaminants that negatively affect the chemistry and quality of surface waters in some areas. Droughts also strain agricultural productivity and could result in increased food prices and food shortages, worsening the situation of those affected by hunger and food insecurity. Ecosystem changes include the migration of disease vectors. The dynamics of disease migration are complex, and temperature is just one factor affecting the distribution of these diseases.

Increased concentrations of ground-level carbon dioxide and longer growing seasons could result in higher pollen production, worsening allergenic and respiratory diseases. Increased carbon dioxide concentrations in sea water may cause oceans to become more acidic and is likely to contribute to adverse ecosystem changes in the Mediterranean Sea and in the world's tropical oceans. This would have potentially dramatic implications for fisheries and the food supply in certain regions of the world. The direct risks of extreme weather events include drowning from floods, injuries from floods and structural collapse. Indirect risks outnumber the direct risks and will likely be more costly. Potential indirect effects include aggravation of chronic diseases due to interruptions in health care services, significant mental health concerns both from interrupted care and from geographic displacement and socio-economic disruption resulting from population displacement and infrastructure loss.

Sea level rise increases the risk of extreme weather events in coastal areas, threatening critical infrastructure and worsening immediate and chronic health effects. Saltwater entering freshwater drinking supplies is also a concern for these regions, and increased salt content in soil can hinder agricultural activity in coastal areas.

Conclusions

Most of the examples of the effects of climate change in this article have been focused on the Mediterranean. This is an area of high environmental vulnerability according to the recent findings of the Intergovernmental Panel on Climate Change (IPCC) reports (IPCC, 2007). It is an area where the most vulnerable populations (children and the elderly, and particularly those suffering from other causes) are at particular risk. The Mediterranean is threatened not only by anthropogenic and natural extreme climate and weather phenomena. It is vulnerable also to the synergy of such phenomena in the event they coincide. The Mediterranean environment is by its nature fragile and vulnerable (alternating droughts and floods, heatwaves, seismic activity, landslide threats etc.). It is fortunate that the IPCC is now preparing a report on extreme phenomena and their relation to climate change.

That we can avoid such a future is evident from the example of the Montreal Protocol (Zerefos, 2009). I cite this example to show that as the most successful protocol to date, the Montreal Protocol has resulted in a better and more protected environment which was being unintentionally threatened by anthropogenic substances that destroyed the earth's ozone shield against harmful solar UV radiation. A small group of scientists, after the discovery of the ozone hole, pushed the international community to get rid of those substances which threatened the ozone layer (Farman et al., 1985). The mechanisms developed in the process of creating the Montreal Protocol could also be used in the case of a climate agreement, if such an agreement takes place in the future. It is interesting to note here that the lessons learned from the Montreal Protocol have not yet been used in the post-Kyoto agreements. Even the science of ozone–climate interactions has not been fully incorporated into the IPCC scientific reports although a large number of scientists from the ozone community, including the present writer, have been involved in both the ozone assessments and the IPCC assessments in the past two decades.

In light of the above discussion, there is an urgent need to take a new look at previously suggested actions to tackle global warming. I should remind readers that the first weak Kyoto Protocol asked for a 6% reduction in greenhouse gas emissions, relative to 1990 levels; this has now been raised to 20, 30 and even 50% for the year 2050! The question arises: in a global economic and environmental crisis, what is the synergy between the two? Talking with leading economists in Greece and abroad I have found that there is no easy answer to such a question. But at the same time I was told that the economic crisis is expected to worsen the anthropogenic effects in destabilising our climate. The proposal to replace fossil fuels by renewables remains promising. Sequestration and so-called carbon capturing is possible at selected sites, while so-called geo-engineering is controversial. But one thing is globally possible and can be done anywhere on our planet. This is the economical and the smart use of available energy. We need smarter networks, better education and in-depth knowledge of human interference in the environment, beginning with the young and ending with the elderly. Everybody should understand that the earth is a lonely planet and we cannot afford to continue to destroy it.

References

Farman, J. C., Gardiner, B. G., and Shanklin, J. D. (1985). Large losses of total ozone in Antarctica reveal seasonal ClOx/NOx interaction. *Nature*, *315*, 207–10.

IPCC (Intergovernmental Panel on Climate Change). (2007). *Fourth assessment report: Climate change 2007 (AR4)*.

Kuglitsch, F. G., Toreti, A., Xoplaki, E. et al. (2010). Heat wave changes in the eastern Mediterranean since 1960. *Geophysical Research Letters*, *37*, L04802, doi: 10.1029/2009GL041841.

Luterbacher, J., Xoplaki, E., Casty, C. et al. (2006). Mediterranean climate variability over the last centuries: a review. In P. Lionello, P. Malanotte-Rizzoli and R. Boscolo (Eds.), *The Mediterranean climate: An overview of the main characteristics and issues*, 27–148. Amsterdam: Elsevier.

Nastos, P. T., and Zerefos, C. S. (2008). Decadal changes in extreme daily precipitation in Greece. *Advances in Geosciences*, *16*, 55–62.

Nastos, P. T., and Zerefos, C. S. (2009). Spatial and temporal variability of consecutive dry and wet days in Greece. *Atmospheric Research*, *94*(4), 616–28.

UNDP (United Nations Development Programme). (2006). *2006 annual report: Global partnership for development*. New York: United Nations Development Programme.

WHO (World Health Organization). (2004). *The world health report 2004 – changing history*. Geneva: World Health Organization.

Zerefos, C., Contopoulos, G., Skalkeas, G., and Contopoulos, G. (Eds.) (2009). *Twenty years of ozone decline*. Berlin: Springer.

Zerefos, C. S., Kourtidis, K., Melas, D., et al. (2002). Photochemical activity and solar ultraviolet radiation (PAUR) modulation factors: an overview of the project. *Journal of Geophysical Research*, *107*, doi: 10.1029/2000JD000134.

European Energy Policy and Carbon-Free Electricity Generation

Emmanuel Kakaras

Introduction

Today's energy production and consumption are neither sustainable nor efficient from economic, social and environmental perspectives. Without the adoption of immediate and effective measures, greenhouse gas (GHG) emissions will double by 2050. As a result there is a growing awareness worldwide that a portfolio of technologies should be deployed in order to meet the energy challenge. Energy efficiency, carbon capture and storage (CCS), renewable energy sources (RES), nuclear power and transport technologies will all make a major contribution in the progress towards a low-carbon economy as well as towards a secure energy supply.

Industrialised countries are responsible for most of the GHG emissions worldwide and undoubtedly they should lead the effort, as they have both the financial and technical resources. However, developing countries must also support this effort through adjusting their energy policies and promoting new technologies. The European Union has a leading position in resolving energy problems, having set strict targets and regulations as well as developed a number of initiatives to support and align the efforts of the Commission, member states and industry. Additionally, the EU Emissions Trading Scheme (EU ETS) is an important driving force for the realisation of a low-carbon economy.

The European Energy Mix in the Electricity Production Sector and GHG Emissions

The EU's total power capacity in 2007 was 775 MW, with the installed capacity of coal, natural gas and fuel oil power plants totalling around about 430 GW. Large hydro and nuclear power plants represent a significant share of the EU's energy mix. However, since 2000 natural gas and wind energy installations have increased their share significantly, while the capacities of fuel oil, coal and nuclear power plants are decreasing (VGB PowerTech, 2006; Wind Energy – The Facts, 2010).

The fossil fuel deposits worldwide are still sufficient. Specifically, hard coal and lignite reserves are adequate for more than 200 years. However, only the 5% of the total fossil fuel reserves are located in the region of Europe and are comprised mainly of hard coal and lignite. It is projected that they will remain a major source of energy in the near future. In 2030, 70% of the total electricity production worldwide will come from fossil fuels, while in the EU their share will be 60%. The European Commission projects that oil production within EU will be reduced by 76% by 2030, while coal and natural gas production will be reduced by 41% and 59% respectively. As a result, the dependence on imported coal will increase from currently 30% to around 66% in 2030 while for natural gas and oil the dependence will rise to 81% and 88% respectively (VGB PowerTech, 2006; Wind Energy – The Facts, 2010).

The support of RES by the EU has resulted in an increase of the RES share in the total gross electricity production by 60% for the 1990–2007 period. Despite this increased proportion of RES, their contribution to the EU's electricity production remains low, and efforts should be intensified in order to reach the goal for the year 2020, by which time 20% of EU energy consumption must come from renewable resources. In order to achieve the 2020 goal, it is estimated that the EU needs to increase the share of RES in electricity production to 30% (EEA, 2009).

Nuclear power plants nowadays account for about 17% of the total European installed capacity but this is expected to decrease by 2030. Barriers to the expansion of nuclear power include safety concerns and high costs.

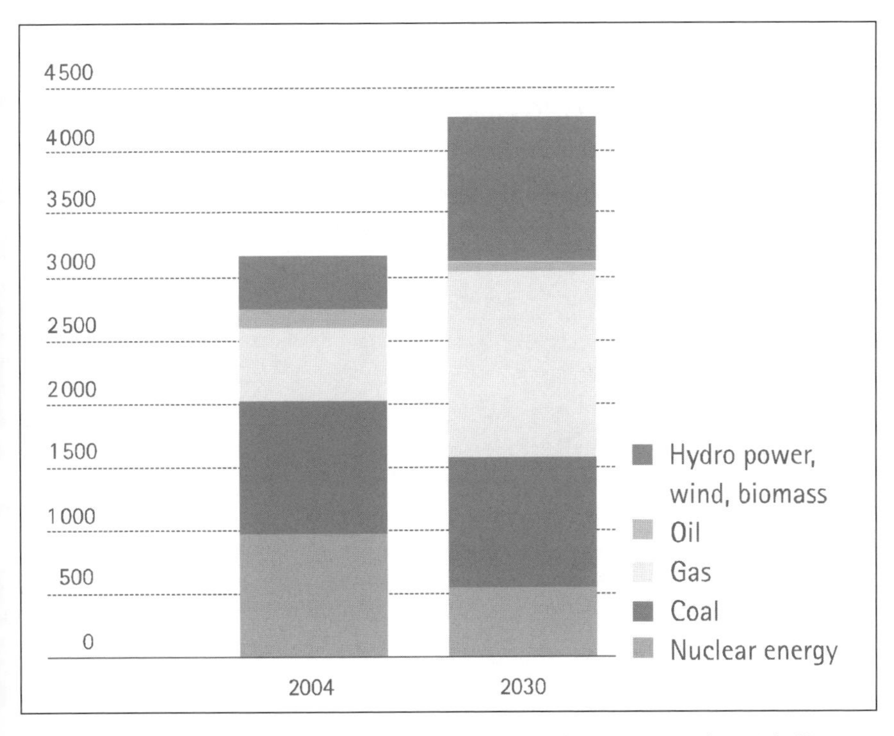

Figure 1. Expected growth in electricity generation in the EU-25 in TWh (Source: VGB PowerTech, 2006)

The EU-27 is responsible for 11–12% of total global GHG emissions, while each EU citizen emits 10.2 t/year CDE (carbon dioxide equivalent), with the world average being approximately 6.7 t/year CDE. These emissions correspond to 473 g CDE per euro of GDP. Fossil fuel combustion is responsible for 83% of total GHG emissions, while 50% of GHGs emitted in the EU-27 is related to electricity production, heat production, road transportation and the manufacturing and construction industrial sector. More specifically, the electricity generation sector accounts for 26% of total EU-27 emissions (EEA, 2009).

According to the European Environment Agency (EEA), for the period 2008–12, the application of measures that have already been adopted by EU-15 member states will result in a 6.8% collective GHG emissions reduction below the Kyoto base-year target (1990). As a result, additional measures should be adopted in order to reach the 8% collective emissions reduction called for by the protocol. According to the EEA, the implementation of the additional measures that have been announced up to now will finally lead to a collective reduction of 8.5% for the period 2008–12 (EEA, 2009).

The EU aims to reduce GHG emissions to at least 20% below 1990 levels by the end of the year 2020. It is estimated that for the EU-27, emissions for 2008 were 10.7% lower than 1990 levels, while the projection for 2020, assuming that all the announced measures are implemented, is 14.3% (EEA, 2009).

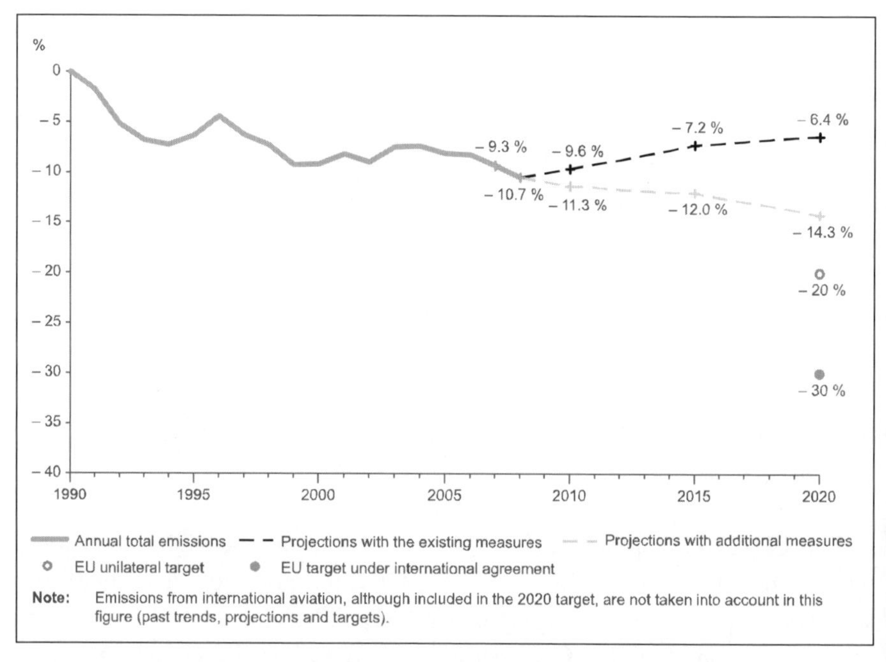

Figure 2. Greenhouse gas emissions reduction for the EU-27 with existing measures (WEM) and with additional measures (WAM) (Source: EEA, 2009)

European Energy Policy

Taking into account the projections that the EU will rely heavily on imported primary energy sources in the future, it is suggested that energy supply security in the EU can be ensured through a sustainable and diverse energy mix, which will render the EU less vulnerable to "energy shortages in imported fuels" and "external energy shocks". Currently, the EU's energy policy focuses on three challenges that have to be faced, namely:

- increasing security of supply;
- ensuring the competitiveness of European economies and the availability of affordable energy; and
- promoting environmental sustainability and combating climate change (COM, 2009).

At the same time, the EU aims to develop a low-carbon economy. In this attempt, no single measure is sufficient. Instead, a combination of measures should be adopted. The energy efficiency involved in energy conversion, energy supply and end-use in an environmentally friendly manner lies at the heart of the problem (COM, 2007). In this context the EU has set targets for 2020 which include:

- a 20% reduction below 1990 levels of GHG emissions;
- 20% of EU energy consumption coming from renewable energy sources; and
- a 20% reduction in primary energy use compared to the projected levels.

To meet these targets, it is of utmost importance for the EU to focus on developing the most promising technologies with the greatest potential. However, assuming that investment in energy innovation does not imply profits in the short term, markets are unlikely to accomplish the goal of a technology breakthrough which will accelerate the realisation of a low-carbon economy unless the appropriate measures (public support) are adopted. Additionally, member states are not able or willing to accelerate the development of those technologies that are not mature enough to be implemented on a large industrial scale, without EU support (COM, 2007).

The European Strategic Energy Technology Plan (SET Plan) aims to stimulate research and encourage the development of low-carbon technologies. According to the Plan, in order to meet the energy targets of 2020, over the next 10 years technology development within the EU should focus on:

- second-generation biofuels;
- CCS demonstration on a large scale;
- off-shore wind power spread;
- large-scale photovoltaic and concentrated solar power commercialisation;
- accomodation by the EU electricity grid of increased electricity produced from renewables and decentralised systems;

- energy efficiency in terms of conversion, supply and end-use; and
- fission technology (COM, 2007).

What is more, the vision of complete decarbonisation by 2050 could be achieved through several technology breakthroughs which need to take place within the next 10 years. These breakthroughs include:

- commercialisation of next-generation renewable energy technologies;
- cost reduction of energy storage technologies;
- hydrogen and fuel cell vehicles commercialisation;
- new generation fission reactors;
- transition strategies towards the development of trans-European energy networks; and
- research in energy efficiency (COM, 2007).

A low-carbon economy gives the EU the opportunity to take a leading position among other global players and have a major share in future low-carbon markets by securing the EU's technology advantage. To this end the European Commission proposed the launching of six European Industrial Initiatives in order to encourage industrial research and "align the efforts of the Community, member states and industry" (COM, 2007). The priorities are the European wind initiative, the solar Europe initiative, the bio-energy Europe initiative, the European carbon capture, transport and storage initiative, the European electricity grid initiative and the sustainable nuclear fission initiative.

During the last few years, EU policy has been the driving force for the implementation of measures in member states aiming for a low-carbon economy. The EU ETS, the Integrated Pollution Prevention and Control (IPPC) Directive, the encouragement to use biofuels and renewables as well as the focus on energy efficiency have encouraged the adoption of new policies at the national level, and EU energy targets for 2020 are expected to intensify this effort. It is projected that by the year 2020 major reductions of GHG emissions will occur mainly in energy and transport sectors (EEA, 2009).

The EU ETS, established through the Emissions Trading Directive 2003/87/EC and coming into force in 2005, refers to large stationary installations (including the electricity generation sector) and covers more than 43% of EU GHG emissions. Each member state, by developing a National Allocation Plan (NAP), determines emission allowances for the installations that are subject to the EU ETS. In the second phase (2008–12), the total verified emissions will be permitted to exceed the total allocated allowances by 10%. In the third trading period (after 2013) the NAPs will be replaced by a single integrated EU mechanism, which will determine the cap for each sector and each installation, while more than 50% of the allowances will be traded (EEA, 2009; Department for Environment, Food and Rural Affairs, 2007).

Carbon Capture and Storage

Given that fossil fuels will be a significant part of the fuel mix for world-wide and European electricity production in the future (70% and 60%, respectively, for 2030) (International Energy Agency, 2007), CCS technology is expected to make a major contribution towards GHG emissions reduction, rendering fossil fuels an environmentally friendly energy source. As a result, EU hard coal and lignite reserves will possibly remain the dominant electricity source in the future, increasing the energy security of the EU and reducing the dependence on imported natural gas.

The CCS European Industrial Initiative proposed by the SET Plan aims at a cost-competitive deployment of CCS after 2020, as well as the acceleration of technology development. The Zero Emissions Platform *CCS EII Implementation Plan* proposes that by 2012 the final investment decision for up to 12 CCS demonstration plants should have been taken. These demonstration plants should be in operation by 2015. In order for CCS technologies to be commercially mature by 2020, the costs of the CO_2 capture process must be decreased through reducing the efficiency penalty implied by their implementation. On the other hand, the storage potential of available geologic formations within the EU (deep saline aquifers, depleted oil and gas fields and unminable coal layers) has to be verified and the storage monitoring techniques and procedures have to be improved. Regarding CO_2 transportation through pipelines, the concept of a trans-European network should be developed and safety and reliability issues should be tackled (Zero Emissions Platform, 2010).

The CO_2 abatement cost for the initial CCS demonstration projects (2015–20) will be from €60–90 per tonne of CO_2. This cost is too high to secure CCS final investment decisions in the near term without any public support. For the early commercial phase (2020–30), the CO_2 abatement cost is expected to range between €35 and €50, while for the mature commercial phase (beyond 2030) the cost will range from €30–45 (McKinsey & Company, 2008).

The investment cost of the large-scale CCS demonstration projects is expected to be €500 million–1,200 million higher than a conventional power plant (without CCS), while operating costs will be dominated mostly by fuel price variation and the price of the ETS allowances (McKinsey & Company, 2008).

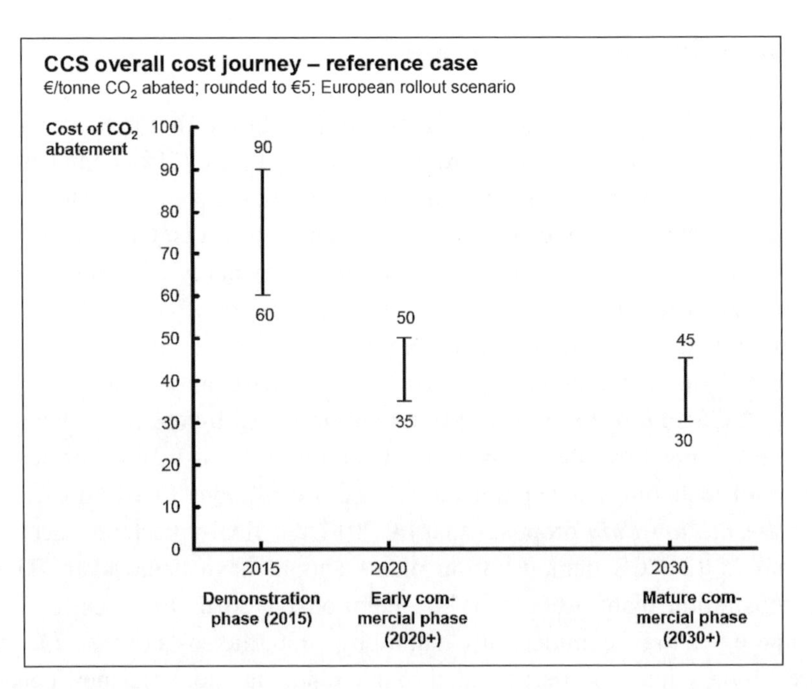

Figure 3. CO_2 abatement costs for CCS application (Source: McKinsey & Company, 2008)

The EU CCS demonstration projects are currently supported through the European Economic Recovery Plan (EERP) and the New Entrant Reserve (NER 300) (Tindale and Tilford, 2010):

- EERP: In 2008, €1 billion was allocated to CCS demonstration projects from the EERP. In December 2009, the European Commission announced the funding of six projects (two pre-combustion, three post-combustion and one oxyfuel) (table 1).
- NER 300: Under the ETS, revenues from the sale of 300 million emissions permits will be used to fund CCS projects by the end of 2011.

Table 1. Funding of six CCS demonstration projects from the EERP

Project	EERP (M€)
Germany, Jaenschwalde	180
Italy, Porto Tolle	100
Netherlands, Rotterdam	180
UK, Hatfield	180
Poland, Belchatow	180
Spain, Compostilla	180

The commercial viability of CCS projects without the need of any financial incentive is related to the EU ETS carbon price. It is forecasted that the EU ETS carbon price for the period 2012–15 will range from €30–48 per tonne of CO_2, while for the period until 2030 the price will be at the same level or slightly increased. In this scenario, CCS technology will be commercially viable by year 2020 (Bellona Europa, 2008).

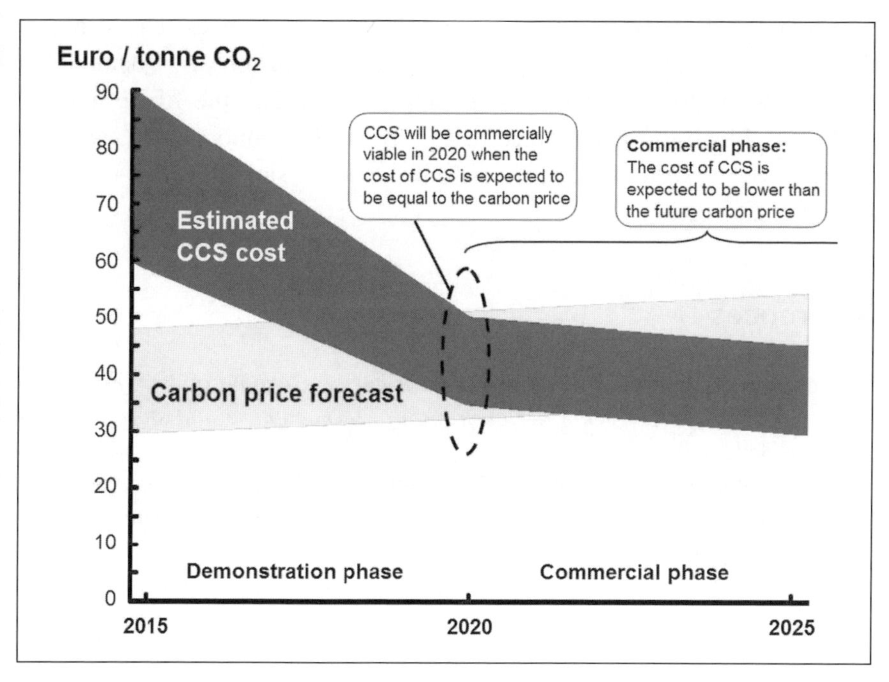

Figure 4. CO_2 abatement costs and EU ETS carbon price evolution (Source: Bellona Europa, 2008)

Conclusions

The EU is obliged to meet multiple energy challenges, namely the security of supply, competitiveness and environmental sustainability. To this end, a combination of measures has been adopted and a portfolio of technologies involved. The EU's ambitious targets for 2020 are supported by adopting regulations such as the IPPC Directive. The EU ETS will also make a major contribution, as will the encouragement of the research and deployment of low-carbon technologies. The SET Plan sets the basis for the technology roadmap needed to fulfil the 2020 European targets and the 2050 vision, while the launching of six industrial initiatives aligns industrial efforts with those of the member states and the Commission.

As fossil fuels are necessary to secure a sufficient supply of energy for the coming decades, and the deposits worldwide – especially coal deposits – are still sufficient, CCS is an effective tool. The cost-competitive deployment of CCS technologies after 2020 lies at the heart of the European CCS policy and is strongly related to the price of ETS allowances. The technical development of CCS will be boosted by the construction of 12 demonstration plants by 2015, which will be supported through the EERP and the NER 300. Finally, it is projected that CCS will be economically viable by 2020, since the CO_2 abatement cost will be at the same level as ETS carbon prices.

References

Bellona Europa. (2008). *Paying for a decent burial: Funding options for an EU programme for full-scale demonstration of CO_2 capture and storage.* Brussels: Bellona Europa.

COM (Commission of the European Communities). (2007). *A European Strategic Energy Technology Plan (SET-Plan): "Towards a low carbon future."* COM (2007) 723 final. Brussels, 22 November.

COM (Commission of the European Communities). (2009). *Investing in the development of low carbon technologies (SET-Plan).* COM (2009) 519 final. Brussels, 7 October.

Department for Environment, Food and Rural Affairs. (2007). *EU Emissions Trading Scheme: Approved Phase II National Allocation Plan 2008–2012.*

EEA (European Environment Agency). (2009). *Greenhouse gas emission trends and projections in Europe 2009: Tracking progress towards Kyoto targets.* European Environment Agency Report No 9.

International Energy Agency. (2007). *World energy outlook: China and India insights.*

McKinsey & Company. (2008). *Carbon capture and storage: Assessing the economics*. McKinsey & Company.

Tindale, S., and Tilford, S. (2010). *Carbon capture and storage: What the EU needs to do*. London: Centre for European Reform.

VGB PowerTech. (2006). *Facts and figures: Electricity generation 2006.*

Wind Energy – The Facts. (2010). Intelligent Energy – Europe. Available at http://www.wind-energy-the-facts.org, accessed 25 October 2010.

Zero Emissions Platform. (2010). *CCS EII Implementation Plan.* European Technology Platform for Zero Emission Fossil Fuel Power Plants.

Targeting the Maritime Dimension of Climate Change: The Role of the European Union's Integrated Maritime Policy*

Antonia Zervaki

The impact of climate change can be observed in the coastal, insular and marine environment in Europe. Extreme weather events such as heat waves and draught, storms, extreme precipitation and the resulting floods are already evident in different European regions. The rise of the Baltic Sea, the floods in Venice and the recent drought in Cyprus are some of the most illustrative examples of the impact of the changing patterns of global climate in Europe. The challenges for European marine ecosystems are significant: temperature changes have an impact on ecological and biological processes, leading to alterations of species' geographic distribution, the extinction of species unable to migrate or adapt in the new conditions and the emergence of new combinations of species. The social, economic and cultural impacts of climate change are equally important; climate change alters the traditional character of economic sectors such as fisheries, agriculture and tourism, while desertification, sea level rise and environmental migration (Bardsley and Hugo, 2010) are expected to change European human geography if preventive action does not take place. In addition, the role of climate change as a "threat multiplier" to international security should not be neglected. Conflict over marine resources, new territorial claims and the opening of new navigation routes due to environmental changes and melting sea ice may alter the global geopolitical status quo (High Representative and European Commission, 2008), affecting European interests and external relations.

This paper addresses the policy tools and initiatives adopted by the European Union under the general rubric of the new integrated maritime poli-

---

* This is an updated, extended version of an article published in *Liberal Emphasis*, *42* (January–February–March 2010).

cy, in order to address climate change and its impacts on the European maritime regions. The European Union's response to the novel challenge of climate change is twofold. On the one hand, the organisation contributes to global efforts under the Kyoto Protocol towards the mitigation of greenhouse gas (GHG) emissions. The EU's climate change legislation covers a series of concrete measures to reach the Union's commitment to reduce emissions to 20% below 1990 levels by 2020. On the other hand, the European Union seeks to enhance its resilience to climate destabilisation through the implementation of adaptation measures as well as initiatives towards the modernisation of the European economy and infrastructure, in order to prevent further human-induced increases in atmospheric concentrations of greenhouse gases.

The European Commission's White Paper on adapting to climate change (COM, 2009a) constitutes the strategic framework for the reduction of the vulnerability of the EU vis-à-vis climate change. The White Paper depicts the main challenges for the marine and coastal ecosystems as well as the social and economic impacts on European maritime regions. It is true that climate change issues in the marine, coastal and insular environments have already been integrated in several sectoral EU policies (environment, fisheries, regional, research and development, transport, external relations). What the White Paper emphasises, though, is the need for a comprehensive approach; thus, the document recognises the fundamental role of the EU's new integrated maritime policy towards this aim.

Climate Change and Maritime Governance in the EU

The publication of the Blue Paper on the EU's integrated maritime policy (COM, 2007) and its adoption by the European Council in 2007 (Council of the European Union, 2007) mark the beginning of a new era in the European Union's approach to maritime governance (Zervaki, 2010). The new policy follows an intersectoral approach aimed at creating a multilevel system of governance of maritime affairs based on the principle of subsidiarity. According to the Blue Paper, the integrated maritime policy "will enhance Europe's capacity to face the challenges of globalisation and competitiveness, climate change, degradation of the marine environment, maritime safety and security, and energy security and sustainability". The new policy is based on "excellence in marine research, technology and innovation" within the framework of the Lisbon Agenda for jobs and growth, and the Gothenburg Agenda for sustainability.

As far as climate change issues are concerned, the new integrated maritime policy is structured as follows: (a) protection and sustainable use of the marine environment; (b) adaptation of fisheries to the emerging challenges and impacts of climate alterations (c) establishment of a zero waste–zero emissions maritime transport policy; (d) enhancement of marine and maritime research; and (e) effective crisis management in cases of natural disasters caused or aggravated by climate change.

The environmental pillar of the new policy, the Marine Strategy Framework Directive (Directive 2008/56/EC), sets the framework for member states to achieve and maintain good environmental status in their marine waters by the year 2020 at the latest. The Directive's principal aim is to enhance the environmental endurance of marine ecosystems in Europe in order to counterbalance the effects of global climate change. Under this direction, member states must define the good environmental status of the maritime zones falling under their jurisdiction,[1] taking into consideration data provided by monitoring, among other things, the impacts of climate change such as biodiversity levels, the intrusion of non-endemic species to local and regional marine ecosystems, the general condition of biological marine resources and the patterns of change of hydrographical conditions (EEA, 2010). The second stage is the adoption by member states of national strategies aimed at the conservation and rehabilitation of their marine environment.[2]

---

[1] The Marine Strategy Framework Directive applies to the marine waters of member states, taking into account the transboundary effects on the quality of the marine environment of third states in the same marine region or subregion. According to the definition provided, "marine waters" are "(a) waters, the seabed and subsoil on the seaward side of the baseline from which the extent of territorial waters is measured extending to the outmost reach of the area where a Member State has and/or exercises jurisdictional rights, in accordance with the UNCLOS, with the exception of waters adjacent to the countries and territories mentioned in Annex II to the Treaty and the French Overseas Departments and Collectivities; and (b) coastal waters as defined by Directive 2000/60/EC, their seabed and their subsoil, in so far as particular aspects of the environmental status of the marine environment are not already addressed through that Directive or other Community legislation."

[2] According to the Marine Strategy Framework Directive, national marine strategies "shall apply an ecosystem-based approach to the management of human activities, ensuring that the collective pressure of such activities is kept within

The European Marine Strategy is complemented by the Water Directive (Directive 2000/60/EC) that establishes the regulatory framework for the protection, apart from inland surface waters and groundwater, of transitional and coastal waters,[3] thus ensuring the sound management of land–sea water interaction in order to prevent and address climate change impacts in a comprehensive way. Last but not least, land–sea interface in the context of managing the impacts of extreme weather phenomena and disasters due to climate destabilisation falls under the scope of the Directive on the assessment and management of flood risks (Directive 2007/60/EC), including the floods from the sea in coastal areas.

The second pillar of the integrated maritime policy, the Common Fisheries Policy (CFP), is closely linked to the Marine Strategy Framework Directive. The fundamental framework of the CFP, Regulation 2371/2002, aims at the conservation, management and sustainable exploitation of living aquatic resources as well as the limitation of the environmental impact of fishing following an ecosystem-based approach. In view of the fisheries policy reform, the European Commission seeks to protect marine biodiversity in order to prevent further deterioration of marine ecosystems by human-induced activities, contributing in this way to the general goal of good environmental status of the marine environment. Moreover, in keeping with the ecosystem-based approach to the fisheries domain in the "Community waters",[4] the EU has adopted measures to protect marine resources in the high seas. More specifically, the adoption of Regulation 1005/2008[5] on illegal, unreported and unregulated (IUU) fishing further reinforces the Community fisheries regime. This also contributes to international efforts towards better global ocean governance and biodiversity conservation

levels compatible with the achievement of good environmental status and that the capacity of marine ecosystems to respond to human-induced changes is not compromised, while enabling the sustainable use of marine goods and services by present and future generations".

[3] According to the Water Directive, "coastal water" means "surface water on the landward side of a line, every point of which is at a distance of one nautical mile on the seaward side from the nearest point of the baseline from which the breadth of territorial waters is measured, extending where appropriate up to the outer limit of transitional waters".

[4] Regulation 2371/2002 applies to "Community waters", defined as "the waters under the sovereignty or jurisdiction of the Member States with the exception of waters adjacent to the territories mentioned in Annex II to the Treaty".

[5] The Regulation entered into force on 1 January 2010.

through the establishment of a comprehensive system for monitoring the legality of catches (either in Community waters or the high seas) imported to the EU by Community and third country vessels.

Sustainable transport is another dimension of the new European maritime governance system which is inextricably linked to the EU's policy towards climate change. The EU Maritime Transport Strategy 2018 sets the long-term objective of "zero waste–zero emissions" maritime transport. The Strategy attempts to establish a coherent and comprehensive approach to reducing GHG emissions from international shipping, combining technical, operational and market-based measures.[6] Although these efforts are presented as an international common venture undertaken by the EU and the international community through the International Maritime Organization (IMO) and the United Nations Framework Convention on Climate Change (UNFCCC), the Strategy mentions that in the absence of progress (as happened with the Copenhagen Conference in December 2009), the EU should make proposals at European level.

As far as the socio-economic impacts of climate change are concerned, the integrated maritime policy provides a general framework for the management of targeted challenges facing the coastal and insular regions of Europe through the existing policy and financial tools of the European regional policy. Some of the priorities of the European regional policy in this domain are the creation and implementation of coastal and management spatial planning, the improvement of existing infrastructure (especially to

---

6 It should be mentioned that the European Parliament had already adopted resolutions highlighting the fact that European maritime policy must play a significant role in combating climate change "through at least three policies: first, the emissions from ships of substances such as CO_2, SO_2 and nitrogen oxide must be drastically reduced; second, emissions trading must be introduced for shipping; third, renewable energies such as wind and solar power must be introduced and promoted for shipping; calls on the Commission to propose legislation to effectively reduce maritime greenhouse gas emissions and calls on the EU to take decisive action to include the maritime sector in international climate conventions; the integration of ships" (European Parliament, 2007). The European Commission, in view of the Copenhagen Conference on climate change, has endorsed this position. In its Communication on a comprehensive climate change agreement in Copenhagen (COM, 2009e), it is mentioned that "the EU has included CO_2 emissions from aviation in its emissions trading system. As regards maritime transport several market-based measures are currently being examined. If no effective global rules to reduce greenhouse gas emissions from this sector can be agreed upon, the EU should agree on its own measures."

address extreme weather conditions such as floods or draught), the promotion of renewable energy sources and the support, through measures of social character, of communities suffering from the impacts of climate change.

Apart from the existing sectoral policies, a new feature of the EU's integrated maritime policy is the introduction of horizontal policy tools which contribute to the coherence of the European maritime governance system and, among other things, to managing the impacts of climate change. The first feature aims at the establishment of comprehensive spatial planning schemes, applied to both coastal (Kay and Alder, 2005) and maritime regions (Suàrez de Vivero, Rodríguez Mateos and Florido del Corral, 2009). Climate change has been integrated as a crucial parameter of spatial planning in the relevant European Commission's Recommendation on Integrated Coastal Zone Management (ICZM) and in the recently published Communication on maritime spatial planning. The ICZM (COM, 2000), based on an integrated, participatory model of planning and management, aims to address an important bio-physical challenge of coastal areas: combining development with the limits of the local environmental carrying capacity. The maritime spatial planning Communication (COM, 2008c) extends the field of application of the principles and aims of the Commission's Recommendation on European marine space. More specifically, maritime spatial planning is presented as a tool to counterbalance the impacts of climate change in not only bio-physical but also socio-economic terms. According to the Communication, "climate change, in particular the rise of sea levels, acidification, increasing water temperatures, and frequency of extreme weather events is likely to cause a shift in economic activities in maritime areas and to alter marine ecosystems. Maritime spatial planning can play an important role in mitigation, by promoting the efficient use of maritime space and renewable energy, and in cost-efficient adaptation to the impact of climate change in maritime areas and coastal waters."

The second horizontal policy tool of the European maritime governance system is related to enhancing knowledge of the seas and oceans and collecting reliable data on their environmental status. Climate change is one of the cross-cutting thematic pillars of the European Strategy for Marine and Maritime Research published in 2008 (COM, 2008a). The new Strategy promotes interdisciplinary research in order to enhance detection and better assessment of the impacts of climate change on the marine and coastal environment. Mitigation of risks is a central theme; however, what is interesting is the reference to the best use of opportunities linked to the impact of climate change in relation to the Arctic Ocean (see the approach adopted in the regional dimension of the integrated maritime policy, below). The establishment of a European Marine Observation and Data Network (COM, 2009b) is expected to improve Europe's data infrastructure, ensuring interoperability of research and data collection mechanisms as well as an interdisciplinary approach in order to support integrated solutions in the different fields of maritime governance.

Addressing Climate Change: Regional and International Perspectives of the European Maritime Governance System

Apart from integrating climate change issues in the sectoral policies of the EU, the European maritime policy attempts to target and mitigate human-induced impacts of climate change in the marine and coastal environment at regional and international levels. Following an ecosystem-based approach, the organisation promotes multilateral and transnational cooperation in the different European sea basins (Juda and Hennessey, 2001); in addition, the EU participates in global efforts to achieve a legally binding comprehensive agreement for the post-2012 era.

In 2008, the European Commission published a Communication on the Arctic Region (COM, 2008b); this document recognises the Arctic marine and land area as a vital and vulnerable component of the earth's bio-physical and climate system. It constitutes the first systematic effort at the EU level to address emerging environmental and geopolitical challenges due to climate change in the region. According to the Intergovernmental Panel on Climate Change (IPCC, 2007), Arctic air temperatures have risen twice as much as has the global average. Coverage of sea ice and land snow in the Arctic has been decreasing steadily at a high pace. Apart from the impacts on the local marine ecosystem and sea levels, this evolution further accelerates global warming mechanisms. In addition, melting sea ice has significant geopolitical implications: it opens new routes to navigation, makes important mineral resources accessible for exploitation and triggers new territorial claims in the region. The Arctic Region Communication sets forth the following policy objectives: (a) protecting and preserving the Arctic in partnership with its population; (b) promoting sustainable use of resources; and (c) contributing to enhanced Arctic multilateral governance.

A Strategy for the Baltic Sea was published in 2009 (COM, 2009c), establishing the first macro-region in the EU[7] and thus creating a new model of regional cooperation. The strategy comprises four policy domains: environment, economy, energy and transport, safety and security. Sustainable development and mitigation of human-induced impacts of climate change as well as adaptation to extreme weather conditions caused by climate change are incorporated into the Strategy's policy objectives. The Strategy for the Baltic Sea follows a project-based approach, since it mainly comprises consortia from EU countries. Thus, tangible results on targeted objectives are expected on a short- or medium-term basis. This will enhance early assessment and promotion of successful scenarios that will serve as good practice, not only in the Baltic region but also in other European sea basins.

The Communication on the integrated maritime policy in the Mediterranean region (COM, 2009f) is the most recently published European Commission document to follow a regional approach. The document mainly focuses on the challenges arising from the Mediterranean's own variable geometry of maritime governance: the Mediterranean basin comprises EU member states, candidate or potential candidate states and third states; it is also characterised by differentiated obligations undertaken by the abovementioned states according to their international or EU commitments.[8] In addition, a large part of the Mediterranean is made up of high seas; this factor hinders states from organising and regulating activities that may have impact on their maritime or coastal zones or on global issues such as climate change. It should be mentioned that the Mediterranean region was characterised as a hot spot in hydrological change (IPCC, 2007), meaning it is vulnerable to flooding and coastal erosion. Moreover, due to its geographical position and features, the Mediterranean will be one of the first

---

7 According to the definition provided by the European Commission, a macroregion is "an area covering a number of administrative regions but with sufficient issues in common to justify a single strategic approach" (COM, 2009c; Samecki, 2009).

8 It should be mentioned that four coastal states – Libya, Syria, Turkey and Israel – have not yet ratified UNCLOS. Although the role of customary law in the case of the international Law of the Sea is fundamental, this reality contributes to the institutional deficit in the region.

regions affected by environmental migration in Europe. The Communication attempts to address these issues by promoting cooperation among coastal states in the region, mainly in the exchange of good practices and technical assistance provided to third countries through the European Neighbourhood Policy. Two policy documents are currently underway concerning the Black Sea region and the Atlantic Ocean.

Last but not least, climate change is incorporated into the international dimension of the new integrated maritime policy. Climate change constitutes one of the main strategic axes of the relevant Communication (COM, 2009d). The document reveals the Commission's intention to contribute to global efforts mainly through developing new sources of energy and techniques for storing CO_2 emissions, reducing human-induced climate change, including emissions from ships, concluding an international agreement on the post-2012 era and supporting developing countries in facing climate change challenges, mainly through initiatives like the Global Climate Change Alliance.

The Commission expressed its commitment to an international maritime governance system based on the rule of law, reaffirming its intention to contribute to global participation in the United Nations Convention on Law of the Sea (UNCLOS) and other international instruments in the maritime field (Oxman, 1996; Kimball, 2001). Moreover, the Communication highlights the existing institutional deficit concerning the protection of biodiversity in the high seas. In this context, it refers to an older proposal of the Commission for an Implementation Agreement under UNCLOS, which could play a key role in filling gaps in the current legal framework, in particular for the establishment of marine protected areas on the high seas. Both positions aim at the reinforcement of international maritime governance which will assist, among other things, global efforts towards the management of climate change.[9] However, there is no reference to environmental migration and its impact on the social and economic geography or the policy propensities at the European and global scale.

[9] For more on the debate concerning the emerging challenges for the Law of the Sea, see Gavouneli, Skourtos and Strati (2006).

Concluding Remarks

The EU's decisions in relation to the management and mitigation of climate change through its maritime governance scheme are inextricably linked to the overall progress of international negotiations. The EU has worked hard towards the adoption of a comprehensive agreement for the post-Kyoto era. However, the efforts to reach this goal, culminating last December during the Copenhagen Conference, fell short of a binding treaty (Doussis, 2010). During negotiations, the EU took up the role of the "honest broker", trying to reach a compromise between the interests of the group of rich states that were not eager commit themselves to significantly reduce GHG emissions, and developing states that would not agree to undertake the cost of limiting the growth of their emissions. Despite its intentions, the EU did not manage to influence the outcome of the deliberations, leaving space for Chinese and US diplomacy to manoeuvre (Afionis, 2010).

Through the Copenhagen Accord, a non–legally binding agreement, governments have engaged at the highest political level; industrialised countries have submitted GHG emissions reduction targets for 2020 and developing countries have submitted action plans for reducing GHG emissions. In January 2010, the Spanish Presidency of the Council and the European Commission, in a joint letter to the UNFCCC, communicated the support of the EU and its 27 member states for the Copenhagen Accord (Presidencia Española, 2010). The letter comprised two pledges on behalf of the EU: (a) its will to move towards a legally binding agreement for climate protection as of 2012; and (b) the reduction of greenhouse gases by 2020[10] and the commitment to reduce emissions by 30% if other industrialised and developing countries contribute with comparable reductions according to their responsibility and capability.

Despite the fact that the European Union recognises the global character of climate change challenges, the pace of international developments may trigger future unilateral initiatives on behalf of the organisation, especially in the maritime field. On the one hand, European leadership may boost developments at the international level; the conclusion of a legally binding agreement that would replace the Kyoto Protocol or the adoption of an UNCLOS Implementation Agreement on biodiversity of the high seas

[10] The European Union's unilateral commitment to reduce emissions by 20% as compared with 1990 levels is already reflected in Community legislation.

could be the positive dimension of European initiatives. On the other hand, unilateral measures may undermine political and institutional processes underway. The example of the Regulation banning single-hull oil tankers from European ports ahead of the date the amendments to the MARPOL Convention were agreed to enter into force by IMO is an illustrative example. It may constitute a unilateral action only as far as the time of implementation of the agreed measures (Boyle, 2006); however, it has caused friction in the international institutional milieu where significant decisions with global impact are adopted. In this context, the adoption of measures concerning GHG emissions in shipping at European level ahead of a comprehensive agreement within the IMO framework may hinder international efforts towards a global agreement in this field. Thus, the EU has to combine leadership in introducing innovative items to the global agenda, while at the same time it should act as a facilitator in negotiations in international fora (Zielonka, 2008).

The EU's attempt to establish a multilevel and multisectoral maritime governance system will contribute to the creation of a more effective framework for the management of climate change challenges. The novelty of the EU's approach to maritime affairs, especially in relation to climate change, lies in the fact that it integrates prevention, management and rehabilitation (where possible). The success of this European venture lies in the effective implementation of relevant regulation by member states on the one hand, while on the other hand coordinated efforts with the rest of the international community remain crucial. Through the international dimension of the new policy, climate change and its impact on the marine environment are incorporated into the agenda of the external relations of the organisation. The upcoming deliberations within the UNFCCC and IMO will determine the future European initiatives and policy measures. The outcome cannot be predicted; however, the European Union is expected to play a decisive role both at diplomatic and operational levels.

References

Afionis, S. (2010). The EU as a negotiator in the international climate change regime. *International Environmental Agreements* (in press).

Bardsley, D. K., and Hugo, G. J. (2010). Migration and climate change: Examining thresholds of change to guide effective adaptation decision-making. *Population and Environment* (in press).

Boyle, A. (2006). EU unilateralism and the Law of the Sea. *The International Journal of Marine and Coastal Law, 21*(1), 15–31.

COM (Commission of the European Communities). (2000). *Communication from the Commission to the Council and the European Parliament, on Integrated Coastal Zone Management: A Strategy for Europe*. COM (2000) 547 final, 27 September.

COM (Commission of the European Communities). (2007). *An integrated maritime policy for the European Union: Communication from the Commission to the European Parliament, the Council, the European Economic and Social Committee and the Committee of Regions*. COM (2007) 575 final, 10 October.

COM (Commission of the European Communities). (2008a). *A European strategy for marine and maritime research: A coherent European Research Area framework in support of the sustainable use of oceans and seas.* Communication from the Commission to the European Parliament, the Council, the European Economic and Social Committee and the Committee of Regions. COM (2008) 534 final, 3 September.

COM (Commission of the European Communities). (2008b). *The European Union and the Arctic Region*. Communication from the Commission to the European Parliament and the Council. COM (2008) 763 final, 20 November.

COM (Commission of the European Communities). (2008c). *Roadmap for maritime spatial planning – Achieving common principles in the EU.* Communication from the Commission. COM (2008) 791 final, 25 November.

COM (Commission of the European Communities). (2009a). *Adapting to climate change: Towards a European framework for action.* White Paper. COM (2009) 147 final, 1 April.

COM (Commission of the European Communities). (2009b). *Building a European marine knowledge infrastructure: Roadmap for a European marine observation and data network.* SEC (2009) 499, 7 April.

COM (Commission of the European Communities). (2009c). *Communication from the Commission to the European Parliament, the Council, the European Economic and Social Committee and the Committee of the Regions concerning the European Union Strategy for the Baltic Sea Region.* COM (2009) 248 final, 10 June.

COM (Commission of the European Communities). (2009d). *Developing the international dimension of the integrated maritime policy of the European Union.* Communication from the Commission to the European Parliament, the Council, the European Economic and Social Committee and the Committee of the Regions. COM (2009) 536 final, 15 October.

COM (Commission of the European Communities). (2009e). *Towards a comprehensive climate change agreement in Copenhagen.* Communication from the Commission to the Council, the European Parliament, the European Economic and Social Committee and the Committee of the Regions. COM (2009) 39 final, 29 January.

COM (Commission of the European Communities). (2009f). *Towards an integrated maritime policy for better governance in the Mediterranean.* Communication from the Commission to the European Parliament and the Council. COM (2009) 466 final, 11 September.

Council of the European Union. (2007). Presidency Conclusions. Brussels European Council, 14 December.

Council Regulation EC 2371/2002 of 20 December 2000 on the conservation and sustainable exploitation of fisheries resources under the Common Fisheries Policy. OJ L 358, 31 December 2002, 59–80.

Council Regulation EC 1005/2008 of 29 September 2008 establishing a Community system to prevent, deter and eliminate illegal, unreported and unregulated fishing, amending Regulations (EEC) No 2847/93, (EC) No 1936/2001 and (EC) No 601/2004 and repealing Regulations (EC) No 1093/94 and (EC) No 1447/1999. OJ L 286, 29 October 2008, 1–32.

Directive 2000/60/EC of the European Parliament and of the Council of 23 October 2000 establishing a framework for Community action in the field of water policy. OJ L 327, 22 December 2000, 1–72.

Directive 2007/60/EC of the European Parliament and of the Council of 23 October 2007 on the assessment and management of flood risks. OJ L 288, 6 November 2007, 27–34.

Directive 2008/56/EC of the European Parliament and of the Council of 17 June 2008 establishing a framework for Community action in the field of environmental policy (Marine Strategy Framework Directive). OJ L 164 25 June 2008, 19–40.

Doussis, E. (2010). Seal the deal: A new approach to climate change (in Greek). In S. Dalis (Ed.), *From Bush to Obama: International relations in a changing world*, 240–56. Athens: Papazissis.

EEA (European Environment Agency). (2010). *EEA signals 2010: Biodiversity, climate change and you.* Copenhagen: EEA.

European Parliament. (2007). European Parliament resolution of 12 July 2007 on a future maritime policy for the European Union: A European vision for the oceans and seas. 2006/2299(INI).

Gavouneli, M., Skourtos, N., and Strati, A. (Eds.). (2006). *Unresolved issues and new challenges to the Law of the Sea: Time before and time after.* Leiden: Martinus Nijhoff.

High Representative and European Commission. (2008). Climate Change and International Security. Paper for the High Representative and the European Commission to the European Council. S113/08, 14 March.

IPCC (Intergovernmental Panel on Climate Change). (2007). 4th Assessment report on climate change 2007 (AR 4). Available at http://www.ipcc.ch/publications_and_data/publications_and_data_reports.htm#1, accessed 10 October 2010.

Juda, L., and Hennessey, T. (2001). Governance profiles and the management of the uses of large marine ecosystems. *Ocean Development and International Law, 32*, 43–69.

Kay, R., and Alder, J. (Eds.). (2005). *Coastal planning and management*. 2nd ed. New York: Taylor and Francis.

Kimball, L. A. (2001). *International ocean governance: Using international law and organisations to manage marine resources sustainably*. Gland: IUCN.

Oxman, B. H. (1996). The rule of law and the United Nations Convention Law of the Sea. *European Journal of International Law, 7*(3), 353–71.

Presidencia Española. (2010). *Formal European Union announcement of support for Copenhagen Agreement and its greenhouse gas reduction objectives*. Available at http://www.eu2010.es/en/documentosynoticias/noticias/ueonunotificacion.html, accessed 9 November 2010.

Samecki, P. (2009). Macro-regional strategies in the European Union: Discussion paper presented by Commissioner Pawel Samecki, Stockholm, 18 September 2009. Available at http://www.interact-eu.net/macro_regional_strategies/283, accessed 15 September 2010.

Suàrez de Vivero, J. L., Rodríguez Mateos, J. C., and Florido del Corral, D. (2009). Geopolitical factors of maritime policies and marine spatial planning: State, regions and geographical planning scope. *Marine Policy, 33*, 624–34.

Zervaki, A. (2010). Multilevel systems of governance: The new integrated maritime policy of the European Union and its position in the international political and institutional landscape. In S. Perrakis (Ed.), *International cooperation at global and regional level: International institutions in motion*. Athens: Ant. N. Sakkoulas Publishers (in Greek, forthcoming).

Zielonka, J. (2008). Europe as a global actor: Empire by example? *International Affairs, 84*(3), 471–84.

Index

F

Printing: Ten Brink, Meppel, The Netherlands
Binding: Stürtz, Würzburg, Germany